THE AMP BOOK

an introductory guide to tube amplifiers.

BY DONALD BROSNAC

Published by

THE BOLD STRUMMER

110C Imperial Avenue Westport, CT 06880

ACKNOWLEDGEMENT

It is surprising how many people are needed to help produce even a small book. Of the many who gave so much of their time, I want to express my gratitude to:

Angela Instruments (Bethesda, MD), Herschel Blankenship (of Schecter), Doug Campbell, Wayne Charvel, Fancy Music (Santa Barbara, CA), Leo Fender, Walter Fuller, House of Guitars (Phoenix, AZ) Doc Kauffman, Jim Kelly, Seth Lover, Frank Lucedo (California Guitar), Mesa Boogie, McCabes of Ojai, Norms Rare Guitars (Reseda), Aspen Pitman, John Quarterman, Red Rhodes, Paul Rivera, Steve Soest, Strings West (Tulsa, OK), Fred Tavaras, Kim Turtinwald, Valley Arts Guitars (Studio City, CA), and the Vox Advocates.

I would like to express special thanks to Vince Basse (formerly with Fender, now with Studiomaster) for always trying to help me find information for my writing. I also wish to thank Twila Carlsen for her indefatigable efforts in editing and typing.

I wish to credit John Peden of Guitar World Magazine for supplying me with photos of some of the vintage amps.

ISBN 0-933224-05-2

THIRTEENTH PRINTING

OTHER BOOKS BY THE AUTHOR

THE STEEL STRING GUITAR, 1973, Panjandrum Press
THE ELECTRIC GUITAR, 1975, Panjandrum Press
TUNING YOUR GUITAR, 1977, Acorn Pub.
SCIENTIFIC GUITAR DESIGN, 1978, Bold Strummer Ltd.
GUITAR ELECTRONICS, 1980, d.B. Music Co.
GUITAR HISTORY, VOL I: FENDER GUITARS, 1984, Bold Strummer Ltd.

THE AMP BOOK

CONTENTS

INTRODUCTION

A guitarist's music is only as good as the sound coming from the amplifier. It is interesting to note, however, that guitarists exhibit a great deal of interest in learning all about guitars, but they give very little attention to the device responsible for the sound — an amplifier. All that time spent searching for that magic sounding guitar is totally wasted if the amplifier used does not enhance the output of the guitar. The guitar makes the signal, but it's the amplifier that creates the sound we hear. A good amp enhances the guitar signal without adding any muffled, tinny, or discordant tonal colorations.

Because every guitar has a personality and sound of its own, many guitarists own several guitars; however, a number of these same guitarists own only one amp. Since amps also have a unique personality and sound, more than one amp is needed in order to create many shades of music. For example, many amps suited for Jimi Hendrix's "Purple Haze", are completely out of their element with Buddy Holly's "Words of Love", or Bill Haley's "Rock Around the Clock".

When a rock guitarist attempts to recreate the sound of rock standards, it can be very frustrating to always sound like a poor imitation. Knowing amps and how to correctly use them is as important as a primo guitar, for you're only as good as your sound.

Different types of music have their own stylistic formats which call for different tonalities. For example, rock music is often a combination of long, singing tones with strong, almost punching notes. The phrase "an iron fist in a velvet glove", might be an apt description. Rock music requires an amp that can give high power levels that range from crystal clarity to raucous distortion. Rock music places the greatest demand on an amplifier. Rock musicians today still prefer the amp sound that was used by Bill Haley, Buddy Holly, and the early Beatles. Early Rock erupted from tube amplifiers which were often pushed to the brink of their operational limits.

The majority of country music has a treble heavy, mildly percussive sound. Amps (similar to rock amps, but non-distorting) with small speakers are used to achieve this type of sound.

Jazz music covers possibly the widest range of tonal variations and requires great versatility from an amplifier. Many jazz musicians prefer their amps to project a very natural tone with no tonal colorations added. For this reason, many prefer solid state amplification.

Blue grass requires a more naturally acoustic sound; therefore it generally uses a microphone linked to a P.A. system.

This volume is being written primarily for rock guitarists and focuses on tube amps; both the vintage models and the newest revivals of tube amplifier technology. This book will cover what an amp is, how it works, what to look for in an amp, highlights of tube amp history, and amps that helped make many guitarists legends.

Tube amps have gained almost mystical renown because there is no thorough book or guide on their history, design or usage. This book is written to help answer the most asked questions and to provide the necessary information so as to enable a guitarist to approach tube amps in an informed way.

Less than 50% of new amps are tube, therefore, this book does concentrate on older tube amps. Because of their popularity, some companies are more thoroughly covered than others.

Keep in mind that this book does not intend to promote an idea that tube amps are always superior to solid state amps. Different amps have different sounds, and the purpose of this book is to help clarify the why and wherefore as to how these tube amp sounds are achieved. But it will make the guitarists aware that if he wants to have a wide range of sounds, he will have to have several amps. This book will help him understand the most effective and economical way to achieve this versatility.

The goal of this book is to provide basic information at an affordable price. To fully cover guitar amplification in depth, it would require a volume many times the size of this book, and the cost of such a work would necessarily be much higher. Since there is more of a need for an affordable guide rather than a consumate reference, the author chose this avenue.

CHAPTER 1

HISTORY OF AMP USE

In early 1904, Fleming invented the world's first electronic tube*; the diode tube, and in 1907 de Forest invented the amplifier tube. These two inventions resulted in the world's first electronic amplifiers. de Forest sold his amplifier tube (the triode) patent to RCA in New York and went to work for them. Coincidentally, in late 1915, Peter Jensen came out with the moving voice coil speaker; this was the beginning of the electrical reproduction of sound. Amplifiers were first used by telephone and radio broadcasting companies. The very first record of amplified speech was on October 21, 1915, at the Panama Pacific Expo in San Francisco, California. It wasn't until the 1920's that amplifiers were specifically made for musical performances. At that time, musicians played or sang into microphones which, in turn, were plugged into amplifiers.

Amplifier design in the 1920's was not specialized. The amps and speakers used for theaters, music, or P.A. systems were virtually the same. The power level of amplifiers was quite low. A 20 watt amp was a fairly large amp. Altec produced a sound system for theaters called the "Voice of the Theater" that was used for theaters holding up to 2,000 or more people. This system had only 20 watts, and was able to fill a hall with adequate volume levels; but only if everyone was quiet. The ratio of loudness vs audience size is something that has made music performances very demanding on sound systems.

All circuits used for tube guitar amps go back to the Western Electric company with branches in Los Angeles, Chicago and New York (the home office in New York was close to RCA). Western Electric was involved with public address and music systems, and in 1927 backed the first talkie, "The Jazz Singer". This was a portent of their future involvement with singers and musicians. RCA, on the other hand, was to become more involved with radio broadcasting.

AT&T (American Telephone & Telegraph, or Bell Telephone) and Western Electric were jointly involved in the design of the circuits used in Altec's Voice of the Theater sound system. Whereas RCA had many patents involved with radio frequency amplification, AT&T and Western Electric had virtually all the patents involved with audio frequency amplification. If anyone wanted to build a musical instrument amplifier, he had to use at least a portion of Western Electric and AT&T's patented circuits. For this reason, amp manufacturers had to secure license rights and put notices on their amps announcing these license rights. This situation continued until about 1954 when the last of the patents lapsed.

In the early years, musicians played instruments close to a microphone, or put microphones inside instruments. By the mid 20's there were passive mixers (made to handle four or so mics) connected to 10 or 20 watt amps powering one or two electromagnetic speakers. At this time, several persons were experimenting with attaching microphone elements directly onto instruments. The reason for this was to obtain a louder, clearer sound. In 1924, Lloyd Loar, the musician head of Gibson, left that company to form the Vivatone Company which made electrified instruments. In 1928, the Amperite Company, in conjunction with Benjamin Miessner, manufactured contact pickups to attach to instruments. However, neither of these companies were successful. It is unknown what company's amplifiers were used with these products.

The reason these early electrified instruments failed to achieve popularity is probably because the A.C. tube (with a cathode sleeve) for amplifiers was not yet fully developed. Early tube amps all had D.C. tubes as well as cathode/filament heaters, directly heated by D.C. batteries. This produced a very cumbersome and basically non-portable system. The A.C. tube finally came into general use in 1928. It was at this time that portable amps became practical.

Although 1929 was a year of electrified instrument experiments by many musicians, among them Alvino Rey, Les Paul,

*In the USA the word tube is used; in the United Kingdom the word valve is used.

De Armond, etc., it was in 1930 that the first practical electric guitar and guitar amplifier took form on the drawing boards of the Electro String Instrument Company. This company, previously known as the RO-PAT-IN Company, was founded by Adolf Rickenbacker (in 1928 the name was changed).

Rickenbacker manufactured and sold the world's first commercially made electric guitar and electric guitar amplifier in 1931. And in 1932 'Doc' Kauffman, along with the Rickenbacker people, developed what was probably the world's first foot operated volume pedal.

These Rickenbacker guitars and amps were popular and sold very well. Shortly to follow was the sincerest form of flattery — imitation. Many companies started to make pickups, but details on their amps have been lost. Not lost, however, is information concerning amps and instruments made by Dobro and National companies. National amps were probably the first widely popular instrument amps. Don Duffy, lead guitarist for the Riders of the Purple Sage, remembers using a National amp: "The amps got so hot you had to place the chassis outside the cabinet for cooling. The amp was a two piece design and was made to operate in this manner."

The mid-thirties was the time that many guitarists became exposed to the playing of Charlie Christian. Charlie used an ES-150 Gibson electric guitar and a Gibson amplifier. Soon Gibson electric guitars and amplifiers were being seen and heard everywhere.

The close of the thirties saw Rickenbacker, Gibson, National, Dobro, Rowe-De Armond and other amps booming out music around the world. The success of these amps was limited, however, by the state of electronics of that time. Loud speakers of this time were bulky devices that used large electromagnets with D.C. power supply transformers. This was the era before compact, simple, permanent magnet speakers, because permanent magnets were just in their infancy.

In 1939 G.B. Jonas received a patent for inventing a new powerful magnet material: alnico. This magnet material made electric guitars highly practical [See Guitar Electronics, by Donald Brosnac] and caused a revolution in loudspeaker design as well as other related amplifier design.

The immediate birth of modern electric guitars and amps was delayed by the occurance of World War II. When it appeared as though the war was to be won, people in the U.S. once again began to give serious attention to music. It was in the mid 40s that many individuals and companies began to make amplifiers. The design and components of this amp were limited to military surplus parts.

In 1943 two friends working outside of Los Angeles, started to work on electronic parts together. These two people were Doc Kauffman and Leo Fender. Doc was a guitarist and Leo was a radio and television electronic technician. By 1946 they formed the K&F Company which made lap steel guitars and small amplifiers. In 1947 Leo wanted to make a large gamble in the form of greatly expanding the business, especially in the area of amplification. The result was that Doc left the partnership, and Leo formed Fender Musical Instrument Company. Fender amps became known for their ruggedness and power. Soon, Fender amps dominated the market.

It needs to be kept in mind that amplifiers' size and output were still quite small. the biggest P.A. systems were about 45 watts. The largest amps used at that time were the 100 watt monsters used by radio broadcasting stations. (Today many radio stations use 50,000 watts or more).

In the 1950s the birth of Rock and Roll placed new complex demands on sound systems. Loud music in larger clubs with more noisy audiences meant bigger amps had to be used. Designing reliable, large amps is not a simple task because as an amp gets more powerful, more heat needs to be disapated. Many musicians of the 30s, 40s, and 50s remember having to remove an amp chassis from an amp cabinet so cool air could circulate around it.

By the mid 50s there were many companies manufacturing a myriad of guitar amplifiers. During this time of escalation of amp wattages, many reliability problems (mainly due to overheating) were experienced. This was the age of Rock and Roll which became personified in the voice of a tube amp with alnico speakers. These tube amps were often driven harder than their design warranted. Musicians had no choice whether or not to use tube amps, because tube amps were all that were available to them.

In the early 1960s, the explosion of pop/rock music resulted in concerts with huge, often very rowdy, audiences. Power requirements of amps and their reliability at those increased levels, reached crisis proportions for some groups. In particular, the concerts of the Beatles were affairs where hearing the music was virtually impossible. At Shea Stadium in 1965, the Beatles had new VOX 100 watt tube amps, instead of their usual VOX 30 watt AC-30 amps. Yet even with this high power amp, few people in the audience could really hear the music. The invention of the transistor (Bell Telephone again) was seen by many as a wonderous panacea for this malady; therefore, many companies (circa 1962) had no compunction about converting from tube to transistor amps.

By 1967 most amplifier manufacturers had changed from tubes to transistors because transistor amps were cheaper, more reliable, and weighed less. Tube amps appeared to be doomed to obsolescence. But at the 1967 Monterey Pop Festival, the popularity of the tube amp sound was revived. Serious rock musicians revealed their preference for tube amps, and at this festival perhaps nobody created more of an impact than Jimi Hendrix. His style of music, using distortion and feedback, revived the use of tube amps, no transistor amps could adequately imitate the Hendrix sound. Among guitarists, Jimi Hendrix is the most admired musician.

Hendrix had picked up on Marshall tube amps when he was playing in England. The hot-rodded 100 watt Marshall amp, when pushed, would easily exceed its 100 watt rating, and Hendrix rarely used only one amp. Hendrix turned the problematic feedback distortion overtones into a style of music.

As the popularity of outdoor concerts increased, colossal amp set-ups became the norm. Guitarists started to make huge stacks out of 100 watt amps, and backdrops composed of hundreds of speakers could be seen. By the end of the 60s, concerts had escalated to needing 5,000 watts of power.

In c.1966, Ampeg introduced its SVT tube amp with 300 watts and capable of being overdriven to momentary peaks of 500 watts. But for most concert musicians, the problems of tube amps (expense, bulkiness, and unreliability) outweighed the positive aspect of preferred sound. During the late 60s, many musical groups and individuals switched to solid state. Even groups like the Beatles used solid state amps on tour, and by 1970 it seemed that tube amps were to become the dinosaurs of electronics; however, this was not to be. Many musicians could not reconcile themselves to the solid state sound. Consequently, the concept of divided, specialized sound systems and amps was developed: (1) plugging a guitar into a small tube amp, (2) positioning a microphone in front of the amp speaker, and (3) connecting the microphone's output to a modern, powerful solid state P.A. amp. This gave the best of both worlds; the preferable tube amp sound, but with the power and reliability of solid state. This method was responsible for the tube amp revival which has continued to this day. In 1972 the Mesa Boogie tube amp debuted; it was miked through a solid state P.A. system and helped reestablish tube amps. Mesa Boogie greatly contributed to a system that has remained popular for the last 10 years.

CHAPTER 2

AMP CONSTRUCTION

An amp is simply an electronic device that gives a large output voltage when a low voltage input is plugged into it. An amplifier does not directly amplify an incoming signal. Rather, an amp uses an incoming signal as a blueprint for shaping power voltage (either A.C. house current or D.C. battery power) into a large facsimile of the guitar's input signal. An amplifier can do this very accurately or inaccurately, it all depends on the competency of the circuit design and the quality of the components. An amplifier processes electronic signals, but it is the speaker attached to the amp which creates the sound we hear. (See chapter on speakers on pg. 19). Because virtually all guitar amps consist of an amp and speaker, the use of the word amplifier in this book will relate to a complete guitar amplifier with speakers.

Fig. 1 Diagram of an amp.

TYPES OF GUITAR AMPLIFIERS

Trying to discern the differences between amps can be a bit intimidating to a novice, but learning a few amp basics is easy. The best place to begin is in the area of appearance. Amps can be constructed to be a complete self-contained unit consisting of an amplifier and a speaker(s) in one cabinet. A combination like this is called a "combo", Combos can be as small as 1 watt battery operated units less than a foot in size, or as large as a 180 watt 100 lb. amp. The most common combos are 5 to 100 watts.

Fig. 2 A combo amp, a Fender Deluxe.

Another type of construction is to have the amplifier chassis in one cabinet, and the speaker(s) in another cabinet. This is called "piggyback" or an amp "stack". Piggyback amps are sometimes as small as 50 watts, or as big as 300 watts.
The amp section is referred to as an "amp head", and the speaker section is referred to as the "speaker cabinet". The reason for this type of construction is three-fold:

1. A divided amp is easier to transport, carry, and store.
2. Having the amp separate from the speaker helps to protect fragile amp components, such as tubes, from destructive speaker vibrations. It also helps protect speakers from the heat given off by the amp. It was really the birth of high power amps that gave rise to piggyback cabinets. The makers of 100 watt tube amps, in particular Vox and Marshall, saw a need for separate enclosures.
3. Having a cabinet devoted soley to speakers meant an enclosure could be designed to maximize speaker efficiency. This is a factor of major significance to bass amps.

Generally, amps of low or medium output power are combos, and medium and high power amps are piggyback.

Fig. 3 A piggyback amp, example shown is a Sound City 200.

Tube vs Solid State

The two types of circuits used in amps are tubes and transistors (solid state). It's interesting to note that in twenty years there has been nothing close to a unanimous decision as to which is better — tubes or solid state. One person can be heard saying, "The concern of solid state vs tubes is passing, new solid state units sound identical to tubes". Another person can be heard saying, "The current boom of manufactures of tube amps stems from the realization that nothing else sounds as good as tubes". Even blind listening tests have not been able to settle the debate.

What is known is that tubes tend to emphasize even harmonics (less harsh to the ear), and standard transistors emphasize odd harmonics. FET transistors emphasize even harmonics, but the slew rate (how fast the device can process a signal) is much faster than tubes. A tube's slow action rounds the corners of notes, and contributes to the warm, soft sound of the tubes. Although there are still technical differences in the response patterns of tubes and transistors, it is likely that in time, true tube sounds will be achieved with solid state.

Tubes and solid state have different positive and negative aspects. Excessive vibration destroys tubes, and excessive heat destroys solid state. Theoretically, solid state could last forever if not exposed to excessive heat. Tubes, on the other hand, will always eventually wear out. Tubes are high voltage/low current devices capable of delivering high voltage peaks, especially when overloaded. Solid state uses low voltage/high current, this results in simplified power supplies and high reliability. Since solid state is low voltage, the highest output for one power transistor is about 25 watts. Compare that to one 6L6 power tube which can put out over 50 watts. Tubes are high impedance in and out, which means they require an output transformer to meet the low impedance of a speaker. The low impedance output of a transistor can be directly connected to a speaker without a transformer.

The one area where solid state remains preferable is with bass amps. Many bass players like the cleaner sound of solid state. Also, the weight factor is important to bass players. Since basses require a great deal of power, a 300 watt tube amp head can weigh 85 lbs, whereas a 300 watt solid state head can weigh only 25 lbs.

Another advantage of solid state is that it works better with P.A. systems. Modern P.A. amps can put out hundreds, even thousands, of watts of reliable, clear sound, yet still be small and light. Musicians today use these positive aspects (power, size, weight) of the solid state P. A. by plugging their guitar into a tube amp and amplifying the tube amp via a P.A. amp. Solid state is too clear and clean to be a guitar amp per se, but it is perfectly suited to faithfully amplify the mellow sound coming from a tube amp.

Amp Classes

There are different ways an amp circuit can be put together, but there are only a few basic configurations. These configurations are commonly denoted as amp classes, and only Class A, Class AB, and Class B circuits are used in guitar amps.

Class A

Class A circuits normally have very low distortion, but relatively low power output. Because full power is always flowing to the tubes, power consumption is high and the amp runs hot. Class A circuits can yield more sustain than other circuits. An example of a Class A amp is a Vox AC 30.

Class B

Class B circuits are about twice as efficient as Class A amps (i.e., they turn A.C. power into current which drives the speakers). Class B circuits are less expensive, but have more distortion than Class A.

Class AB

Most guitar amps casually called Class B are, in fact, Class AB. This AB circuit uses mainly Class B designs (with a bit of Class A). The AB circuit is a compromise which most designers, manufacturers, and musicians feel gives reasonable power, reliability, and all-around economy.

There are circuit Class C, D, G, H, etc., but these are not generally used as guitar amplifiers.

Amplifier Subclasses

There are two subclasses to amps: 1. SINGLE ENDED, and 2. PUSH-PULL. Each has its own merits. Generally, single ended are used as inexpensive, low-power amps; e.g., the Champ.

Single Ended

This circuit uses a tube which functions as a single stage of gain boost. Several single ended tubes can be joined in series, but the complete circuit is still single ended. See the Fender Champ schematic.

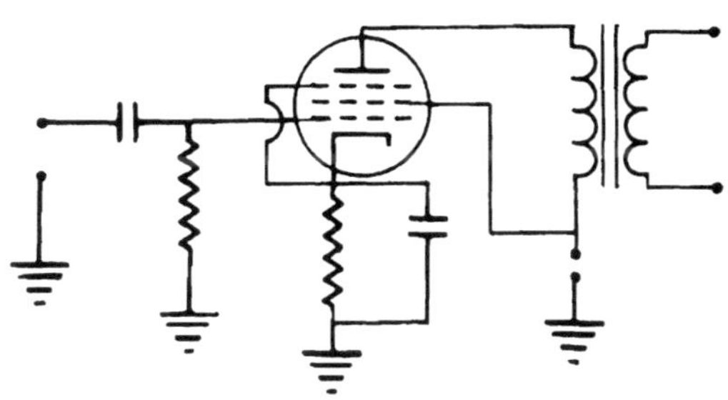

Fig. 4 Single-ended output stage.

Push-Pull

A push-pull amplifier circuit uses two amplifier tubes (or transistors) and can provide a large power output. Push-pull amplifier operation can be likened to a see-saw — one tube is always doing the opposite of the other.

Push-pull circuit operation is as follows: The secondary winding of input transformers is tapped at its center. This center wire is routed to the cathode of the tubes, usually through a resistor and capacitor. The outer ends of the secondary are then connected to the grids of the tubes.

When the winding to number one tube's grid is pushing out a positive signal, the winding going to the number two tube's grid is pushing out a negative signal. When all raw power supplied to the cathode of one tube is being utilized, there is virtually no need for power to the other tube and vice versa. These out-of-step outputs from the two tubes are put back in step in the same manner as the input signal was split: a center tapped transformer winding. The transformer in this case is the output transformer. This means a push-pull amplifier circuit utilizes its power more efficiently because it isn't resting half the time like a single tube amplifier. A push-pull amp can have about twice the output of a single tube amp, but at the same power consumption level.

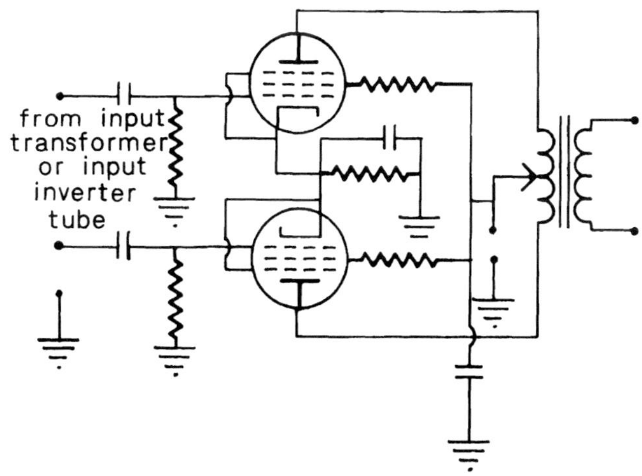

Fig. 5 Push-pull output stage.

Amp Operation

In simple terms, amplifiers take the small voltage signal from a guitar and use it as a blueprint to reshape high voltage A.C. current that the amp plugs into. To accomplish this, an amp has many parts and can be overwhelming in apparent complexity. For simplification, all amps can be broken down into three precise sections: 1. PRE-AMP, 2. POWER AMP, and 3. POWER SUPPLY. The only difference between amps is what kind of and how many components are used.

The Pre Amp

The pre-amp receive the extremely small output of a guitar and increases it to the point where it is recognizable and useable to a power tube, Most of the volume and tone controls are located in the pre-amp sections; e.g., the tone controls of a Fender amp are between the first and second stage of gain.

The 12AX7 is the most common pre-amp tube. Also used are 7025 tubes which were originally a less noisy military version of a 12AX7. Today, a 7025 is thought of as a higher quality tube than a 12AX7.

Since a 12AX7/7025 is a dual tube, the output of one side can be fed into the other. This is called cascading and can be used to increase output and sustain, as well as shaping tonal color.

The Power Amp

This is the real workhorse of an amp, it produces the high wattage to power speakers. It is the most expensive part of a tube amp and it consumes the most power; therefore, it gives off the most heat.

Of all the many configurations of power amps, the most common arrangement is a pair of 6L6GC tubes working as a push-pull; but Vox uses EL-34 tubes. Hi-Watt has used KT88/6550, and Gibson has used 7027. There are about a dozen different power tubes that have been used in guitar amps. Power tubes are not usually cascaded because the output from the first would overload the second.

A power amp often has no controls, but some sophisticated models do have input sensitivity and output level controls.

The Power Supply

The power supply takes A.C. house current and changes it into the many different high voltages needed by a tube amp; e.g., 450 D.C. plate volts, 400 D.C. screen volts, 12 A.C. volts to heater filaments, etc. Current entering an amp at the power transformer is changed to the needed voltages by transformer windings (see illustration). The rectifier, either tube or solid-state, change the A.C. current to D.C. The filter capacitors help smooth the power by removing the ripples in the rectified current.

Schematics

A pictorial drawing of an amplifier appears to be easily understood, but they do have two serious shortcomings: 1. Pictorial drawings can't show all parts of a circuit; wires often cross, and small resistors may hide behind another component, etc. 2. A pictorial drawing doesn't show how a circuit operates; therefore, it is very difficult to trace an input all the way through to the output.

Schematics, on the other hand, offer greater clarity and present a circuit in logical stages. Schematics use symbols which depict the functions of components, this makes it easier to compare circuits. Schematic circuit diagrams almost always have the input on the left side and output on the right, this simplifies following each step of the circuit.

Reading and understanding schematics is vital for anyone doing in-depth work with amplifiers.

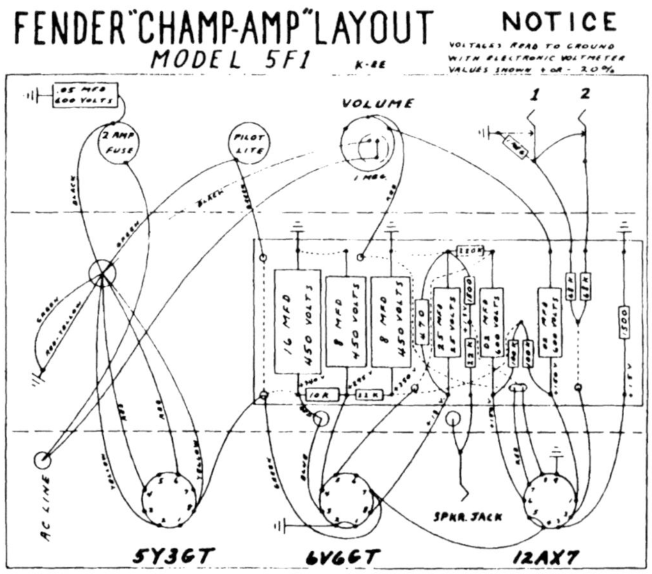

Fig. 6 Pictoral of an amp's circuit.

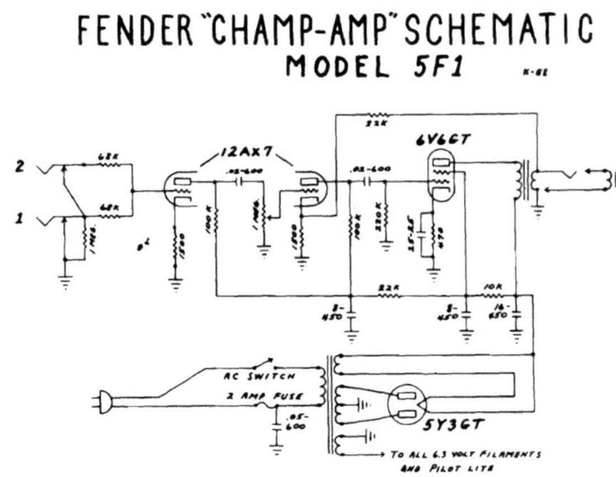

Fig. 7 Schematic of an amp's circuit.

AMP CONSTRUCTION

Why An Amp Is Laid Out Like It Is

If you look at an amp chassis, you may wonder if placement of components really does make any difference. Yes, component placement definitely does make a difference. Generally, an amp will be laid out in a line so that the input goes in one end through the amp's length, and then comes out the other end. However, transformers are mounted at right angles to another; i.e., they are not in line. If transformers are in line next to each other, they can generate hum and noise; mounting at right angles can reduce this problem.

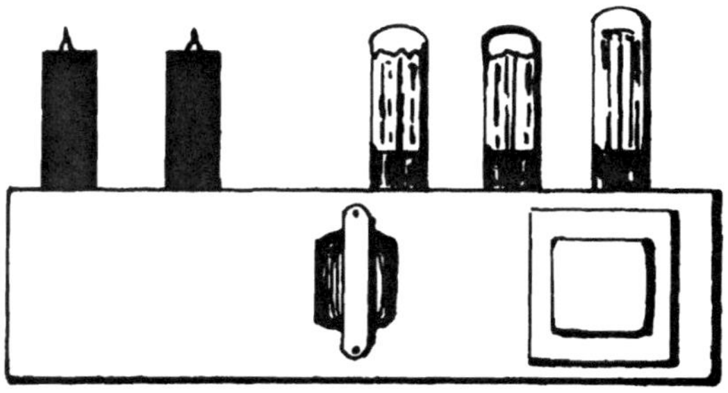

Fig. 8 Diagram of amp layout,

Amp Components

TUBES

The following describes the construction of vacuum tubes used in amplifiers:

A vacuum tube's construction is similar to a light bulb — it has a filament surrounded by a vacuum atmosphere. If you stretch a length of very thin, hard metal wire inside an airless glass bulb and apply an alternating current to it, the flow of current through the wire will cause it to glow. But if there is air inside the bulb, the wire will burn up very quickly. This is the principle of the light bulb, and the first stage of the creation of a vacuum tube.

When a light bulb is turned on, electrons on the hot filament become agitated and break away, streaming into space. If you insert a metal plate inside the bulb (near the heated wire), you could detect the electrons hitting it. The filament emitting electrons is called the cathode, and always has a negative polarity. The plate which is hit by the electrons is called the anode, and it always has a positive polarity. Electric current flows from negative to positive (not positive to negative). There will be no charge received by the plate when the A.C. current is subjecting the filament to positive voltage. A vacuum bulb with A.C. current in, produces D.C. current out (at the plate). This operational mode is called RECTIFIER.

Many modern day tubes, especially amplifier tubes, use separate filament heaters and electron sources. In normal usage, there is a heater behind a filament cathode. The heater heats the cathode so electrons can leave its surface more freely. The use of a separate heater and cathode reduces tube electronic circuit noise and provides for longer tube life. Since separate filaments are so common, most schematics don't show them. The only clue is a note near the drawing of the power transformer which reads: "to all 6.3 volt filaments".

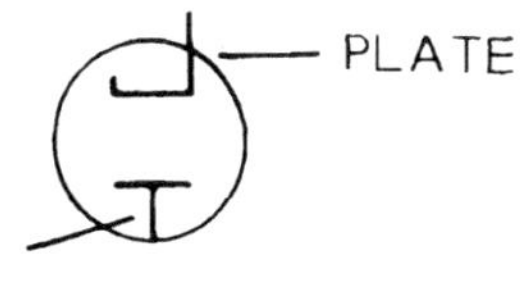

Fig. 9 Diagram of a rectifier.

Note that the D.C. output is a series of pulses. This type of rectifier is called a half wave rectifier. Power for electronic circuits needs to be smoother; therefore, two rectifiers can be put together with their output put linked out-of-phase so that both waves of the A.C. current can be utilized. This is called a full wave rectifier and produces a D.C. current of 120 pulses per second.

If the D.C. current from a rectifier is used without a filter, there will be a tremendously loud tone at either 60 or 120 cycles per second pulses; filters eliminate hum. A power supply filter is composed of either large capacitors, or large capacitors and inductors. The inductive and capacitance can smooth out the D.C by storing energy while the voltage is rising, and releasing it when the voltage falls.

Since a rectifier tube handles a large amount of power, but are somewhat inefficient, a lot of power is lost as heat. This means that rectifier tubes must be large and have sufficient space around them for air circulation.

The Basic Amp tube — The Triode

The heart of an amp is the amplifier tube, and the simplest amp tube is called a triode. Tri means 3, and ode means elements. A triode tube is composed of three elements inside a glass tube with a vacuum inside.

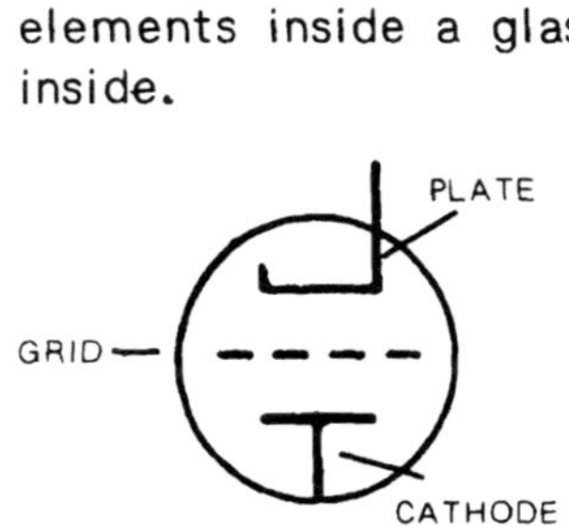

Fig. 10 Diagram of a triode.

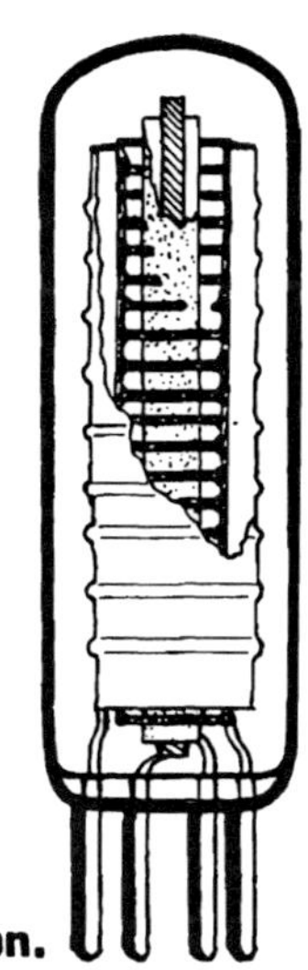

Fig. 11 Actual triode construction.

Though diagrams of tubes show flat grids, screens and plates, tubes are nearly always built in concentric circles. The energy flows from the center out.

A rectifier only has a steady continuous effect on the current it handles, whereas amplifier tubes have an extra element — a grid, which allows for modulation of current. If a screen wire grid is put between the cathode and plate of a rectifier, nothing will happen. The electrons will simply flow from cathode to anode (plate). But if a negative voltage is given to the grid, the flow of electrons will be repulsed. The voltage to a grid is called grid bias. There is a point where the negative grid voltage will be so strong that it completely blocks the cathode to anode electron flow. This is called plate current cutoff or cutoff.

The following is an example of simple amplification: The cathode receives current and tries to send this current to the anode, but it has to go through the grid (connected to the low output signal of a guitar). This guitar signal will cause the grid to act as a valve which turns on and off the current flow from cathode to anode. The reason a guitar's small signal can have such an influence on electron flow is because the grid is much closer to the cathode than to the anode. This makes the grid very sensitive to the guitar signal it receives. The grid only has to be about one-fiftieth as strong as the anode to cathode flow to exert its valve-like action. The actual guitar signal received by the grid is produced as follows: When a guitar string vibrates it is sensed by the guitar's pickup. Downward motion of the string usually creates a positive voltage, and upward motion creates its opposite (usually a negative voltage). If a string vibrates at 80 cycles per second (c#) this will induce an alternating current of 800 cycles per second in the pickup. When this 800 Hz signal is fed to the grid, the grid modulates the cathode to plate (anode) flow. This results in the cathode to anode flow having 800 pulses per second; i.e., the small signal from the guitar shaped the large voltage from the amplifier. The speakers receive the voltage output from the anode.

Though this is the basic priciple of an amplifier, they are more complex because more than one amplifier tube is used. For example, there are additional components (i.e., volume and tone controls) to regulate the sound, and resistors to help make the sound clearer. There is also an output transformer which changes a tube's high impedance output into the low impedance necessary to drive a loudspeaker.

The Tetrode Tube

The name is derived from the Greek tetr (four) and ode (way). A tetrode is a higher performance tube than the triode previously described. Because the capacitance of the cathode, grid and anode limit its performance, engineers developed the tetrode, which has an additional internal element (screen grid) that reduces these interelectrode capacitances from a triode's approximate capacitance of 2 pf to the tetrode's usual .05 pf. This screen grid is placed between the control grid and anode, and it serves as an electrostatic shield. A tetrode is a smoother operating amplifier tube and it is much more efficient. It can amplify a signal up to 26 times more than a three element triode.

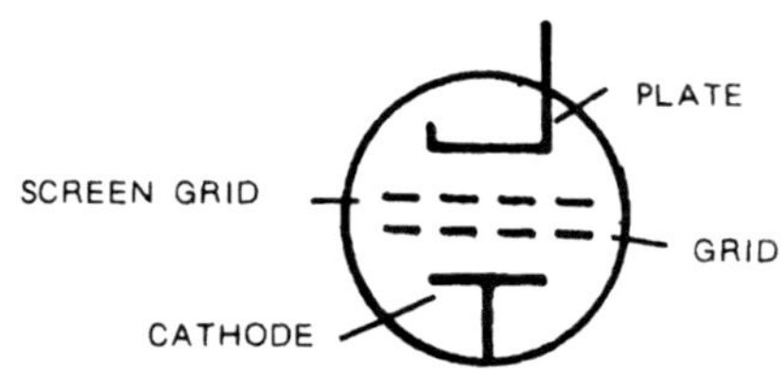

Fig. 12 Diagram of a tetrode.

The Pentode (Five Element)

The pentode is a five element tube which has all the features of a tetrode, plus another component called a Suppressor Grid. This Suppressor connects with the cathode and improves tube performance over that of the tetrode.

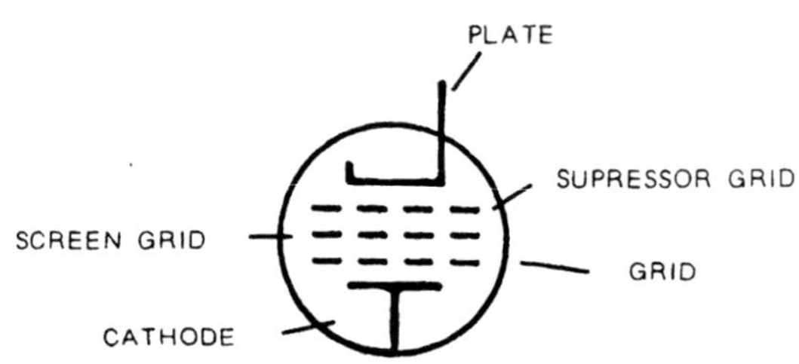

Fig. 13 Diagram of a pentode.

Electrons can be dislodged from the anode when electrons from the cathode strike it. This is called "secondary emission". It is not a serious problem in triode tubes, but in a tetrode, secondary emission can interfere with screen grid operation causing a reverse current flow between the anode and screen grid. A pentode, therefore, has the advantages of a tetrode without its disadvantages.

Multi-Unit Tubes

Multi-unit tubes consist of the internal structure of two or more tubes inside one glass casing. The most common multi-unit tubes are diodes, triodes, and tetrodes. Multi-unit tubes are sometimes called dual tubes. The 12AX7 is a dual tetrode, and is the most common pre-amp tube.

Fig. 14 A dual triode, example is a 7025 Model.

Selection of Tubes

Tubes, like guitars, can sometimes look the same, but sound very different. Tubes made in Europe tend to be of lighter construction than USA tubes; this yields a more fragile tube, but more output and a smoother sound. It's not possible to generalize about the tubes from one country because there are several tube companies in one area. Genalux, Mulland, and Phillips tubes come from England; Telefunken tubes from Germany, and GE and Sylvania (mostly) from the U.S. Every company has their own way of building a particular tube. Superiority of one over the other is dubious because different designs yield different sounds. Even two tubes of the same model from the same amp company can vary from each other. A manufacturer may supply a batch of one tube designation (e.g., 6L6), but some may be completely U.S. made, whereas others might be partially made in Mexico or Brazil. Perhaps the most baffling tube aspect is that each tube made at the same plant, on the same day, may sound different.

Inside the upper portion of most tubes is an area that looks chrome plated. When a tube is new, the glass is completely clear. But when the manufacturer tests a tube by heating it, berrylium inside the tube vaporizes which causes the chrome-like appearance. The vaporized berrylium counteracts the oxidation of tube operation and extends tube life. As a tube wears out, you begin to see ghost rings around the inside. When a tube is completely worn out from oxidation (more than the berrylium can counteract), a tube will turn completely cloudy/white inside. Tubes often stop working because of broken or shorted internal components; this usually occurs before complete oxidation is encountered.

Pre-amp tubes are usually of poorer quality than power tubes. In the U.S., GE and Sylvania are now making an upgraded version of the 6L6GC called the STR 415. This is the tube used by Mesa Boogie. Anyone with an amp that uses a 6L6 can benefit from the STR 415 because it is more reliable and gives more output.

Matched Tubes

Matched tubes have matched electronic responses. Tubes can vary, e.g., two unmatched Sylvania 6L6GC tubes can have diferent output levels even if they have the same input. Matching output transistors is accepted as necessary for proper solid-state amp performance; however, this same theory is not universally accepted with tube amps. Most tube amps have a provision (usually a trimmer potentiometer to adjust grid bias) to match tubes, but this requires the work of a serviceman. Many newer amps have self-matching circuits for power tubes.

The Groove Tube Company is the major company currently selling matched tubes. Aspen Pitman, the founder of Groove Tubes, is almost single-handedly responsible for making everyone conscious of matched tubes. Aspen feels that mismatched tubes yield unpleasant distortion, shorter tube life, and lower output. Although the aspect of a more pleasing sound from matched tubes is debatable, matched tubes can last longer and give a little more output. Groove Tubes are more expensive because every Groove Tube is hand tested. Purchasing Groove Tubes can reduce the frustration of buying tubes only to find they are defective and must be returned. Considering the value of time and money, pre-tested and matched tubes can be a more practical investment.

Fig. 15 Groove Tubes.

Tube Names — American Standard

Tubes have names which are composed of either numbers or letters with numbers. Typical tybe names are 6L6, 12AY7, 6BQ5, and 5U4. The first number (6, 12, or 5) denotes the voltage of the tubes filament heater. A 6L6 is a six volt tube, a 12AY7 is a tweleve volt tube, and a 5U4 is a five volt tube.

If there are two letters in the middle of the code, it denotes the evolution of the tube. A 12AY7 was devised after the 12AT7, but before the 12AX7. If there is only one letter, it can relate to the tube's evolution, or it can be totally arbitrary.

The second number space denotes the number of elements in the tube, e.g., a 5U4 has four elements (two plates and two filaments) whereas a 5Y3 only has three elements. A 12AT7 has seven elements (this denotes it as a dual triode), and a 6BQ5 has five elements (it's a pentode).

One or two letters after the main code name denote extra features of that tube. The original 6L6 was a metal case tube, and when the first 6L6 glass tube was made, it received the name 6L6-G. The G was for glass. When this tube was improved, it was called the 6L6-GA; later versions became the 6L6-GB and 6L6-GC. The 6L6-GC is the most common power tube.

Tube Names — European Standard

European tubes use their own code; some common models are the ECC-81 and EL-34. The E stands for European design. The L stands for pentode, and C stands for triode. An ECC tube is a dual triode. The ECC-81 is a more advanced (higher gain/lower noise) version of a 12AT7, and was designed after the American 12AT7 came out. After the ECC-81 debuted, the 12AT7 was upgraded as the 12AT7A — a tube even better than the ECC-81. Similarly, an ECC-82 has been upgraded from the 12AU7, and the ECC-83 upgraded from a 12AX7.

There are both European and American tubes available for the same basic configurations. In addition to the ECC-81 and other dual triodes, there is the EL-34 power tube, an overseas version of the 6CA7.

Special Tubes

Not conforming to standard tube nomenclature are tubes which are generally considered higher grade versions of a common tube. A 7025 is a high gain/low noise version of a 12AX7, and a 7048 is a special 6V6. Note that these tubes are designated by numbers only. The four digit code of many special tubes origianted when these devices were created for military/government usage.

The KT series of tubes, e.g., the KT66 and KT77 are of either American or European design. These tubes are custom versions of EL-34 (KT77) and the 6L6 (KT66).

Common Guitar Amp Tubes

Interstage pre-amp tubes

These tubes are used in stages preceding the power tubes, they can be used for preamps, reverb send and receive, tremolo, or inverters. Nearly all tubes of this classification are triodes. They could be used as power output tubes, but their output would only be one to two and a half watts. For this reason, interstage tubes do not have output ratings, but rather, they have ratings relating to gain. The following are the most common interstage tubes.

12AT7 — Originally developed for radio frequency ECC81 amplification.

12AU7/ECC-82 — This tube has less gain than several other interstage tubes; but one unique feature is that it is lower impedance and, therefore, better for lower impedance circuits.

12AY7 — A low noise tube (developed before the 7025).

7025/ECC-83 — A 12AY7 type tube having very low hum and noise.

Interstage tubes often are enclosed in a

metal sleeve to shield against spurious noise such as that from TV and radio.

Power Tubes

Power tubes can be worked with a variety of voltages, the higher the voltages to the plate and screen, the higher the output wattage possible For instance, using a 6L6-GC with 360 volts to the plate and 270 volts to the screen, could yield an output of 26.5 watts (for two push-pull tubes). Using 450 plate volts and 400 screen volts could yield 55 watts; and 500 plate volts and 500 screen volts could yield 75 watts output. The reason that all amps don't use the highest plate and screen voltages possible is that high voltages require very heavy duty (and expensive) power transformers; also, tube life is shortened when output wattages are raised. It is important to remember this — one amp with four 6L6GCs could have an output of 53 watts and another could have an output of 150 watts. Just because two amps use the same tubes, doesn't mean that the power or sound of the amps will be the same.

6AQ5 — This tube is one of the oldest power tube pentodes. The low output of this tube has relegated it to use in very small amps.

6BQ5 - EL84 — This tube is an advanced version of the 6AQ5. It is used in many amps, especially Vox AC-15 and AC-30 amps.

6V6 — This is the standard tube in small American amps with a push-pull circuit. Sometimes (as in the Champ) it is used singly. Note: Fender uses special heavy duty 6V6 tubes.

6L6-5881-KT66-STR415 — The 6L6 is the standard tube for nearly all medium or high power American amps. The 5881 is an advanced 6L6 with a unique base. The 5881 was supplanted by the 6L6-GC. The KT66 is an alternate design to a 6L6-G. The STR-415 is a special heavy duty higher performance version of the 6L6-GC. Note: Fender has tubes custom made to its specs and their 6L6-GC is comparable to a STR-415.

6550 - KT88 — These tubes with close cathode control grid spacing make them somewhat susceptible to shorting-out from vibration. This tube is used in many European amps such as Marshall.

7027 — This fairly heavy duty, high power tube has an awkward base design which greatly limited its use.

Tube Data — Class AB1 Power Tube Output
(with two tubes in a push-pull circuit)

Tube	Plate Volts	Screen Volts	Power Output Watts
6AQ5A	250	250	10
6BA5	250/300	250/300	11/17
6V6 - 7408 (6V6GTA)	285/400	285/400	14/30
6L6 - 5881-KT66-STR415	360/450	270/400	26.5/55
7027	450/540	350/400	50/76
6CA7-EL34-KT77			
6550-KT88	400/700	275/500	55/100

Note: Tubes can be run at many voltages. Higher voltages produce higher outputs

Rectifier Tubes

Rectifier tubes waste a lot of power, give off great amounts of excess heat, and are expensive to use because they periodically wear out. In 1982 Fender completed their switch-over in all their tube amps to solid state rectifiers. The most common tube rectfiers were the 5Y3, a three ampere unit. The 5U4 is a four ampere unit. Amps that have tube rectifiers can use plug-in solid state units, but only if the components of the amps are able to withstand the higher outputs of the solid state device.

Transformers

The most simple tube amps (i.e., a Fender Champ) use two transformers: a power transformer and an output transformer. A push-pull amp will need a third transformer (center-tapped) for use as a phase inverter if a tube is not used for this purpose. Also, a spring reverb amp will need two small audio type transformers to change impedance to and from the reverb unit.

Transformers are used to change voltages and impedances. Power transformers take A.C. house current of 110 watts and provide (via coil taps) the voltages needed to power filament heaters (6.3 and 12.6 volts) and high voltages to the plates of the tubes (450 volts). Output transformers take the high impedance output of the power tubes and turn it into low impedance suitable for the speakers. Since power transformers handle much more voltage than output transformers, they are larger in size. Heavy duty transformers can supply more power to the tubes.

The performance of a transformer influences the sound an amp will yield. Some people prefer the compressed and distorted tone of an amp which is achieved by using an overextended transformer, e.g., some Marshall amps. Many English amps and some old Fender amps give a quality of distortion often called grinding. This happens when the amp's signal overloads the output transformer and causes a condition called transformer core saturation. The bifilar transformers in Marshall amps contribute to the unique Marshall sound. It is possible to fit this type of transformer into other amps if an English grinding tone is desired. Fender amps, especially newer models, have very robust transformers and the compression and distortion obtained is from circuits other than overload/overworked transformers.

The general color code for power transformers is as follows: Two black leads are primary input; two red leads are high voltage (or secondary) out; one red and yellow wire is the center tap of high voltage out; two yellow wires go to the rectifier filaments, and any other pair of green, brown, or gray wires are low voltage filament secondaries. The general color code for output transformers is: blue and red wires are primaries from the output tubes, and green and black wires are the secondary going to the speakers.

The following are specifications for a transformer:

1. Electrical specs of primary and secondary windings.
2. Single or dual primaries (for running on two voltages — e.g., USA vs Europe.
3. Voltages you want secondaries to carry.
4. Turn ratio of windings.
5. Amount of shielding.
6. Core design, e.g., "C" core or "H" core.

Resistors

Resistors are the most common part in any electronic circuit. They are used to create and divide different voltages. The small resistors are made of carbon. Carbon is semi-conductive so it can be made into configurations of various conductive/resisting values. Common carbon resistors are cheap, but they add a bit of spurious noise (due to electron motion and microphonic action) to a circuit. Deposited carbon film resistors are more quiet, but much more expensive.

In addition to carbon resistors, which are limited to handling just a few watts of power, amps also use wire wound ballast resistors. These are wound over a high temperature form, and can take many watts of power. These resistors help to dissipate heat and power in the power supply section of an amp when the filaments do not have a step-down transformer, and the combined resistance of tubes almost equals the necessary resistances for voltage drop. Ballast resistors fill out the remaining needed load.

Capacitors

Capacitors are electronic devices which have two functions in an amplifier: they store a charge or block low frequencies so as to allow high frequencies to pass. Capacitors are rated in values of capacitance called "farads". Very small capacitors are now rated in pico farad (pf). The old term was micro-micro farad (uuf). Older amps might designate uuf caps instead of pf. Micro farad or uf = 1,000,000th of a farad; .01 uf is the same as 10,000pf; and .0001uf is the same as 100pf.

There are five basic types of capacitors in amps: 1) paper, 2) ceramic, 3) mica, 4) electrolytic, and 5) mylar.

Paper Capacitors

These inexpensive caps use rolled paper as a dielectric and were the industry standard. Paper capacitors are composed of two long sheets of foil separated by wax or oil impregnated paper. One of the foil sheets becomes the outside and acts as a shield to the inner windings. For this reason, these caps have a band printed on them to indicate the lead connected to the outer layer. The wires coming out of a paper cap connect to these parts. The whole unit is dipped in wax to seal it. Paper caps are cheap and were the standard capacitor for years, but since about 1960, companies have been using ceramic and mylar caps because they are more stable. Paper capacitors break down more often than other capacitors. But if not exposed to excess humidity, heat etc., they can last more than 20 years. Paper caps tend to be somewhat noisy and microphonic. They are typically inaccurate. A 16 MFD paper capacitor could be between 14 to 25 MFD. Checking a cap's specifications before using is not very helpful because they tend to change value after use.

Ceramic Caps

Because ceramic disk caps are cheap, durable, and of small size for their capacity, they are now the most common capacitor.

Mica Caps

Mica caps use a piece of mica as the dielectric. Mica capacitors are very stable, have good efficiency with low loss, and are used when the frequencies are high and losses are to be kept low. Mica caps were used a great deal in precision parts of an amplifier circuit. Mica caps have a color code to specify capacity. There are three color code combinations for mica caps. All codes give the value in pico-Farads. The numerical value for a color is the same as the resistor color code.

The first three dots (the upper row with the arrows) give single numbers, and the right bottom dot is the multiplier. This example would be 610 pf. The dot on the bottom left indicates voltage rating. The middle bottom dot indicates tolerance value. The third code for caps is the outdated American War Standard.

Electrolytic Capacitors

Electrolytic capacitors are made by separating two pieces of metal foil in a paste. These are used when high values of capacitance are needed and are typically used as filter capacitors in power supplies and as cathode bypass capacitors. These capacitors have polarity, i.e., current can only flow one way through them. These polarities are indicated by + or - on the casing. Standard electrolytics are available from 1uf to thousands of uf with working voltages of 3 to 700 volts.

Mylar Caps

Mylar film caps are high quality caps that are used if the need for quality outweighs the higher cost of a disc. Caution: Rapid discharge of a cap can cause electrical shock. A 1mfd cap charged with 500 volts can store 25 watts of power.

Diodes

Diodes used in place of tube rectifiers are two part solid state devices. They are more electrically efficient than rectifier tubes. Today the word diode is generally reserved for referring to a solid-state diode rectifier. Diode tubes are now generally referred to as rectifier tubes. Solid-state diodes are much more sensitive to heat than tube rectifiers.

Circuit Boards

A circuit board is used as a mounting place for all small components. Early amps had resistors and caps wired into a vulcanized fiber board plate. In the 1960s many manufacturers began to use etched (printed) circuit boards. The best grade is G10; lower grades have more paper in the composition. Minute fractures can ruin the circuit, be very hard to find, and difficult to repair.

Tube Sockets

Tube sockets are devices that tubes plug into, and the best tube sockets are mica filled (a durable insulation material). Cheaper plastic insulation filler can breakdown around the pin sleeves; consequently, the tube can burn out. When soldering a pin sleeve, be sure that no solder drips into the the sleeve. This can interfere with the plug connection.

Chassis

The metal box holding all the parts of the amplifier is called the chassis. Nearly all chassis today are steel. This is often chrome plated for durability. Steel chassis

Fig. 16 An amp chassis, example shown is a mid-1950's Fender Deluxe.

are strong, but the softer aluminum chassis have some positive points.

Aluminum chassis can make a quieter circuit since it is non-magnetic and transformer fields would not be conducted. Also, aluminum can help dissipate heat. The amps made by the Danelectro company had aluminum chassis.

Amp Circuits Providing Special Effects

The only control the first tube amps had was a volume knob. To inhance tone shaping, treble, mid-range and bass controls were added in the forties and fifties. Musicians and audiences liked more variety in sounds. This led to special sound effects circuits.

The first special effect was rythmic fluctuation of volume. The name for this effect is tremolo and it first appeared on amps in the fifties. It is a simple effect to add to an amp. A more complicated effect to build into an amp is vibrato, the rythmic rising and falling of pitch. This also appeared in the fifties. Another effect, which became common in the sixties, is reverb. Reverb is the creation of an artificial echo.

More advanced effects are those of phasing/flanging, chorus, compression, expansion switchable channels and graphic equalization. These effects began to be built into amps in the seventies and eighties.

Tremolo and vibrato are often incorrectly used terms. Tremolo is common, but true vibrato is very rare. The Magnatone amps with vibrato have true vibrato. Other amps carrying the term vibro often don't have vibrato. Fender has been lax in the correct use of terms. No Fender amp has true vibrato, but true vibrato is available by using the "Tremolo" bridge on Fender Stratocasters.

Reverb is the creation of an artificial echo. Amps with the inclusion of a reverb circuit generally have one knob on the control panel to adjust the amount of reverb (echo).

Although there are both mechanical and electronic reverb circuits, virtually all tube amps use mechanical reverb, which consists of long springs stretched the length of a narrow box. An electrical signal from the amp is changed (via a transducer) from an electrical signal into a physical motion of the spring; motion is changed back to an electrical signal (via another transducer) and this is routed back into the amp circuit. When a signal travels the length of a spring, the sound in effect literally bounces back and forth. This is the creation of echo.

Because spring reverbs need transducers and are basically inefficient, a set of reverb springs (with transducers) in a box can not simply be plugged into an amp to achieve reverb. An amp with mechanical reverb must have several additions: 1) an extra tube to boost the signal going into a reverb unit, 2) a transformer to match the impedance of the reverb input transducer, 3) a transformer to match the impedance of the reverb output transducer with the input of a tube, 4) an extra tube to boost the signal coming from a reverb unit, and 5) a control to blend non-reverb sound with reverb sound so that gradations of reverb can be achieved.

Reverb units usually have two or three springs. Two long springs were standard for years, but now three shorter springs are favored because shorter springs are more controllable and less prone to feedback due to bumps to the unit and vibrations caused by high sound levels.

Fig. 17 Reverb box in cover.

Fig. 18 Inside reverb box showing springs.

CHAPTER 3

AMP SPEAKERS

How A Speaker Works

The function of a speaker is to produce sound waves by moving air. The function of an amp is to create a strong enough electrical signal that will stimulate movement of the speaker's voice coil. A cone is attached to this voice coil, and when the voice coil moves, the cone moves. It is the cone's back and forth motion which moves air, and therefore, creates sound.

The first device to actually reproduce sound was not a speaker, but rather, a device called an acoustic horn. These horns were connected to the membrane diaphragm of a record player, and were the first sound reproduction devices (see diagram). Acoustic horns were scientifically developed and even average ones could fill a living room with sound levels far above those of pleasant listening. The efforts given to the building of acoustic horns reached their peak in such complex designs as folded horns capable of both good volume and strong bass response.

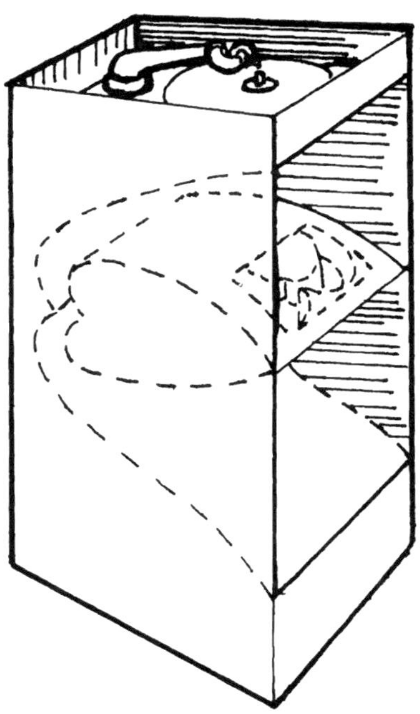

Fig. 19 A complicated acoustic horn.

In 1915 when Peter Jensen invented the moving voice coil speaker, vast possibilities for sound reproduction were born. Adjustment of treble/bass tone response and much higher sound levels were achieved. These first loudspeakers had the voice coil winding (which was connected to an amplifier's output) surrounded by an electrical magnet field coil; hence, the name "field coil speaker". The first field coil speakers used a variable field coil and a lever-actuated diaphragm. Soon after, a cone diaphragm speaker with a directly attached voice coil was developed; this speaker became the standard speaker until the late 1940's.

The A.C. output of an amp flowed through a voice coil, and as the sound wave reproduction current changed, the voice coil would be either attracted to or repulsed by the electromagnetic field coil. The speaker cone attached to the voice coil would then be moved back and forth creating sound waves in the air.

Fig. 20 A field coil speaker in a 1930's amp.

These field coil speakers had some problems. The electromagnetic field coil required high voltage of D.C. current. At first, batteries were used, but batteries were short-lived, heavy, and expensive. For this reason, speaker power supplies were constructed. This was achieved with a diode which converted 120 volt A.C. house current to D.C. This current would then go through a transformer to be transformed into the higher voltage needed to power the electromagnetic coil.

Another problem was that transformers

and diodes often emitted a great deal of hum which was picked up by the antenna. Though humbucking transformers were developed in time, field coil speakers were too bothersome for almost everyone.

The possibility of a light-weight, trouble-free speaker came about in 1939 with the development of the permanent magnetic material — alnico. Before this, the idea of using a permanent magnet in a speaker (instead of an electromagnetic field coil) would have meant a huge, expensive, and very heavy magnet. But alnico changed all that because an alnico magnet could be less than 1/10 the size of an earlier tungstun magnet and still exert more force.

The only problem with alnico in 1939 was it was a very expensive material, and this was during the time of the Great Depression. In 1941 when World War II began, all available alnico was diverted into the war effort, and it wasn't until after the war ended in 1945 that alnico became available to private industry. The amplifiers made by Gibson, Fender, National, Rickenbacker and others utilized field coil speakers until the late forties. In fact, "permanent magnet" speakers using Alnico magnets didn't become popular until the early fifties.

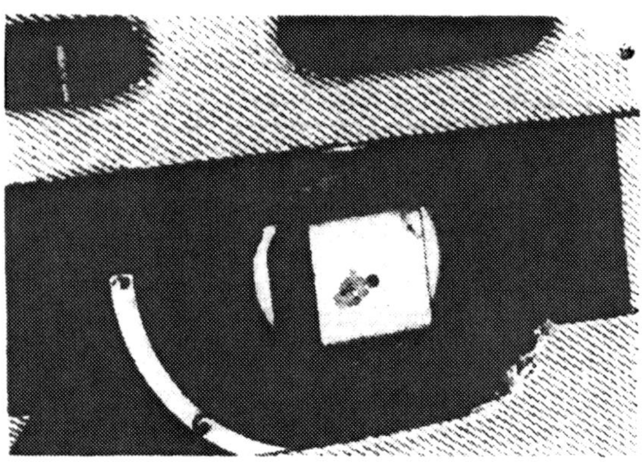

Fig. 21 An alnico magnet speaker.

The standard magnet material for speakers was always alnico V. The other alnicos, such as alnico II or III were never felt to offer any benefits over alnico V. During the ten years of the 1950's, alnico speakers became so ordinary that few people thought of changing them. Since alnico worked fine, there was no reason to look for other materials. But in 1960 trouble was brewing for alnico. The main suppliers of the cobalt used in alnico (countries such as Zaire) began to raise the price. Soon, alnico was becoming exceedingly expensive. As a result, new magnet materials were developed. The standard magnet material for the 60s, 70s and 80s has been those composed of oxides in a ceramic base. Ceramic magnets, plentiful and quite economical, have now become the standard.

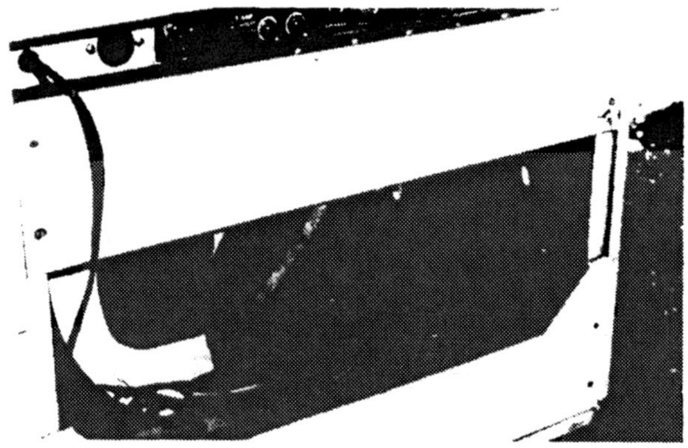

Fig. 22 A ceramic magnet speaker.

A speaker's output ratings throughout the years reflect the desired amp output and volume levels popular at that time. This is because speakers have always been designed to match amplifiers.

The following information charts the evolution of the 12" guitar amp speaker.

1930/39 — 35 to 40 watts, field coil design
1940/49 — 35 to 40 watts, first alnico V speakers
1950/59 — 100 watts, speakers begin to fully exploit alnico's power
1960/69 — 100 watts, new glues for voice coils
1970/79 — 150 watts, ceramic magnets better glues, and aluminum wire voice coils became more common
1980/19-- 300 watts, more advanced ceramic magnet designs with large vented voice coils.

Loudspeaker Construction

The materials that go into a speaker, and how those materials are crafted, all influence a speaker's performance and the final sound of an amp. A top notch guitar and amp are of little worth with a speaker, even a good one, that is not matched to a musician's needs. For this reason, an examination of speaker construction is very worthwhile.

Speaker Frames

A speaker is held together by a speaker frame, sometimes called a basket. The frame's purpose is to keep all speaker elements in the

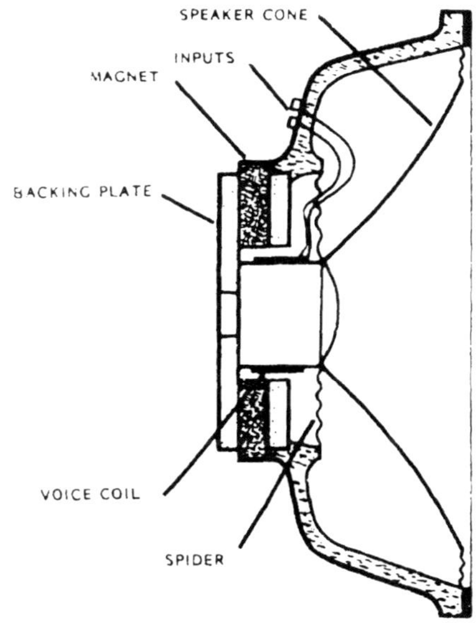

Fig. 23 Parts of a modern loudspeaker.

proper geometry. Frames can be of steel, aluminum, or some special alloy such as magnesium. Frames stamped out of sheet metal are the least expensive, whereas exotic alloy frames are the most expensive.

The frame centers the cone, both at the surround and at the spider. It also is the mounting base for the voice coil gap/magnet base. A frame must be rigid enough so it doesn't bend or flex from cone motion, magnet assembly, weight, or voice coil heat dissipation. If a frame fails at this job, the voice coil will eventually rub in the gap and fall apart.

Many speaker frames for guitar amps are inexpensive steel stampings. Most, however, are well-made and quite stiff. The expensive alloy frames on some speakers are also stiff, and have an important advantage over steel — they carry heat away from the voice coil area. The arms of such a speaker frame behave like cooling fins. Virtually all speakers capable of handling high wattage inputs have alloy frames.

Speaker Surround

A speaker cone's outside edge is supported by the hoop of the frame. The oldest speakers had the cone's paper edge directly attached to the frame. While this does provide secure mounting and good core alignment, it does limit cone motion which, in turn, wastes power. Also, this type of mounting makes a cone stiffer, and less able to produce low frequency tones. For these reasons, speakers have an added band around the cone to hold it in place. This band is called the surround, it is located between the cone and the frame. Cone surrounds also serve the function of helping to dampen vibration reflected by the rim of a speaker frame. This dampening action results in reduced erratic tones from phase cancellations.

The subject of surrounds may sound trivial, but it is thought of as one of the most important parts of speaker design. The proper stiffness/compliance of a surround determines the efficiency and durability of a speaker. Changing a surround can radically alter a speaker's behavior; e.g., with a treated paper edge surround, a speaker's lowest note may be 75 Hz, but if a more flexible cloth surround is used, that speaker may be able to produce a note as low 45 Hz. Changing surrounds in this manner can turn a guitar speaker into a bass speaker. The treated edge is the most basic surround, it is formed by extending and corragating the edge of a cone and then applying a gum-like material to the surround surface. The gum material makes the surround more limber so the core can move more easily. It also makes the surround less brittle.

A second type of surround is the cloth surround , sometimes called "cambric". A cloth surround is usually not as strong as a treated edge, but it does facilitate easier cone motion.

The third type of surround are rubber surrounds, composed of a strip of solid, thin (about 1/32") rubber approximately 1" wide encircling the edge of a cone. Rubber surrounds are quite pliable, yet tough. Since they are pliable, they can have a problem keeping a cone centered; therefore, they are of limited used in guitar amps.

The fourth type is the foam surround which uses a strip of foam around the cone. Foam surrounds are very pliable, allow easy cone motion, and are not suited to be guitar amp speakers.

Of all types of surrounds, the most common used in guitar amp speakers are the treated edge and cloth edge.

Speaker Cones

With all the space-age electronic technology of the twentieth century, it may come as a bit of a surprise to find that speaker cones made in the 1980s are made of the same material as those made in 1915 — paper. Although many companies (e.g., JBL) have experimented with other materials, such as aluminum and plastics, almost everyone returns to paper. True, alloys and plastics are used in many high frequency tweeters, but for full range amp speakers no material possess significant advantages over paper for cones. Paper for cones is not simple paper, it has certain specifications engineered into it such as rigidity, weight, and thickness.

Rigidity

If the paper pulp is pressed very hard, a stiff, thin, non-porous cone is formed. A cone such as this would be more effective in forming high notes than lower notes. If less pressure is utilized, a thicker, more pliable cone is produced which yields fuller bass notes.

Weight

The amount and weight of the paper pulp and bonding glue controls the weight of the cone. The heavier the cone, the lower notes it can produce. But heavier cones take more watts to drive, and a heavy, fast moving cone can be difficult to keep aligned.

Thickness

This should not be confused with density or rigidity because a cone can be thick yet porus. In general, thick cones tend to distort less and give a flatter frequency response.

Cone Size

The size of speaker cones has a great deal to do with the tone of a speaker. Small speakers are generally treble heavy and large speakers can excell in bass. Many guitarists consider four 10" or two 12" speakers right for guitars, and 15" and 18" speakers the choice for bass. Although size is often indicative of frequency response, it should not be used as a sole means to judge the sound of a speaker. A high quality 4" speaker with a 1½" voice coil and 20 oz. magnet could give much more bass than an economy 12" speaker with 3/4" voice coil and an 8 oz. magnet. Generally, cheap speakers are weak in bass response.

Another factor of cone size is how tones are projected from the cone surface. The smaller the cone, the more it projects its sound straight out from its surface. For instance, a 4" speaker may sound very loud while you are directly in front of it, but if you step to the side, you'll notice a drop in volume. A speaker that directs its sound only straight ahead is not useful when the audience is spread out. Likewise, a speaker that fans out its sound far to the side is wasting watts of power if there is no audience. Speakers 18" or more in diameter tend to be too diffused, and speakers under 8" tend to be too directional. Below is a list of speakers and at what frequencies they tend to become unidirectional.

18" — 800Hz
15" — 1,000Hz
12" — 1,300Hz
10" — 1,600Hz
8" — 2,000Hz
4" — 4,000Hz

The angle of a speaker's cone shape influences the dispertion of a speaker's sound. The deeper the cone angle, the more directional it will be, and the more it will assist treble tones.

Voice Coil

The voice coil is composed of small diameter wire wound around a tubular form. The two output wires of an amp are connected to the two ends of the voice coil winding. The wave fluctuations in the amp's output attract and repel the voice coil to the speaker's magnet. A secondary action of the current to a voice coil is the production of heat. The more output wattage a speaker voice coil receives, the hotter it gets; and if it gets too hot, it can disintegrate.

The earliest speakers had voice coils of of only 1/2" in diameter, and could not handle high wattage. But at this time it wasn't necessary because amps were small and did not have high wattage. As amps increased their wattage, voice coils became larger so as to handle the high wattage. Larger voice coils can dissipate heat more easily. In a test, it was determined that in just a few hours with a 200 watt input consisting of a smooth test sign wave (see diagram) the voice coil of a 12" speaker could reach over 500°. Rock music of 200 watts, however, is much rougher than a 200 watt test wave. In rock concerts, voice coils sometimes glow red hot, and metal speaker frames become too hot to touch. Larger voice coils can handle more watts because the heat is spread out over a wider area. Standard voice coils can handle the following wattages: 1/2" = 5 watts, 1" = 30 watts, 2" = 100 watts, 4" = 200 watts.

Several things can be done to increase voice coil strength: 1) to help dissipate heat, the windings can be made from aluminum wire instead of copper, 2) a vent in the front of the voice coil (the dome) or in the back of the magnet structure can assist cooling, 3) special glues can be used to hold the voice coil together.

Magnet Structure

The magnet of a speaker is the vital part of a speaker. A large magnet is generally desired, but there is much more to magnets than size and weight. The most important magnetic criteria is the flux density in the voice coil gap. The stronger the magnetic flux, the stronger the dampening of the cone/voice coil motion. This means that if a speaker has a stronger flux field it can handle more watts of power before voice coil motion becomes uncontrollably destructive. This is the reason there is so much hoopla about the size and strength of a magnet.

Speakers that used alnico magnets had an easy time with flux fields, but a ceramic magnet's flux is more difficult to focus. This is because ceramics have a much lower flux density than alnico. Sometimes as much as three times the amount of ceramic magnet mass is needed to equal an alnico magnet. This situation resulted in a redesign of speakers. Alnico speakers generally had the magnet inside the voice coil, but ceramic magnets of equal power are too big to be inside a voice coil. Nearly all present day guitar speakers have ceramic magnets held by iron or steel plates; this assembly surrounds the voice coil.

As for differences in the tonal response of alnico speakers compared to ceramic, few people hear any difference. Regardless, ceramics have the possibility of clearer sound. Alnico tends to act slower and produces blurred tones when fast single note tones are played.

Speaker Spiders

The term "spider" is given to the part of a speaker that centers the voice coil in the voice coil gap. The earliest spiders were made out of 1/32" phenolic (linen impregnated with resin), and their intricate shape lead to the nickname — spider. Although the early spiders were an open pierced design (for flexibility). Contemporary spiders are made of solid, stiff cloth. A spider must keep the voice centered even when it's moving back and forth a quarter of an inch (when loud bass notes are sounded).

The spider must keep the voice coil moving in the gap without striking the magnet structure. The Gauss Speaker Company has developed a dual spider system to achieve this. The drawing shows how this works. Whether or not two spiders are necessary is debatable.

EFFECT OF SURROUND, CONE AND SPIDER

The behavior of the moving parts of a speaker contribute tonal qualities. If the above elements are heavy, a speaker could be capable of powerful low notes. Similarly, if these elements are pliable, it will be easier for a speaker to produce low notes. Limber cones will flex too much at high frequencies, and this flexing will impede the creation of a clear high treble. If a speaker is to generate tones in the treble range, then qualities of light weight and stiffness must be designed. Below is a graphic depiction of how stiffness and weight effect oscillation speed of a steel ball on a spring. This corresponds to speaker action. Note that stiffness and light weight would create a high frequency motion and vice-versa.

SPEAKER CABINETS

The purpose of a speaker cabinet is to prevent the sound waves emitted from the front of a speaker from being sucked back toward the rear of the speaker cone (see diagram). A cabinet separates the waves from the front and rear of a speaker. The notes with long wave lengths (low notes) would especially be lost. The size and shape of speaker cabinets does influence the type of sound achieved. There are five basic speaker designs: open back, closed back, front loaded horn, rear loaded horn, and bass reflex.

Open Back

This is the type of construction of most amp/speaker combos (a combo is a unit with speaker and amp in one cabinet.) Open back construction limits bass response to some degree, but it does yield a more brilliant sound.

Fig. 24 Open-back speaker cabinet.

Closed Back

This construction yields a strong, solid sound with a full bass. It is criticized by some as being a bit boomy with unclear reverberation.

Fig. 25 Closed-back speaker cabinet.

Front Loaded Horn

This construction is good for projecting mid-range and bass tones. For this reason, it it is often used for bass speaker enclosures.

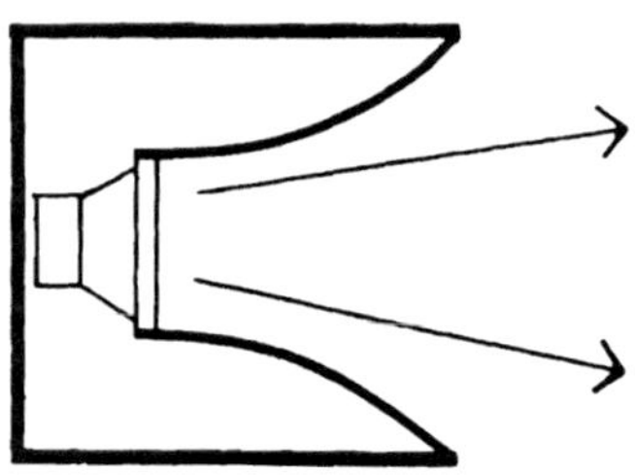

Fig. 26 Front loaded horn.

Rear Folded Horn

This is most effective for strong bass tones, but less suitable for higher frequencies. quencies.

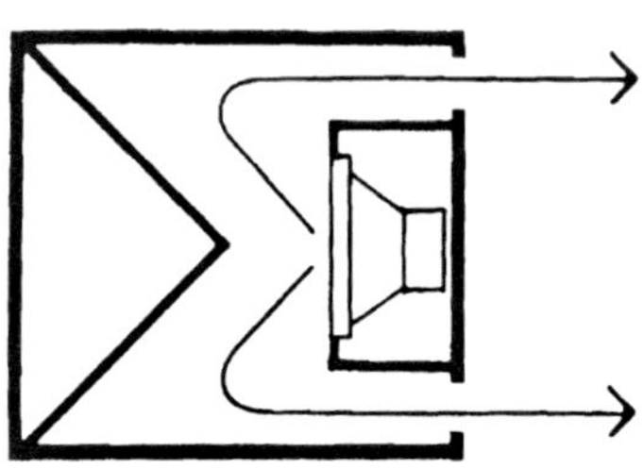

Fig. 27 Rear Folded horn.

Bass Reflex

This cabinet can provide a high volume, low bass response, but in order to be efficient, it requires a large size cabinet.

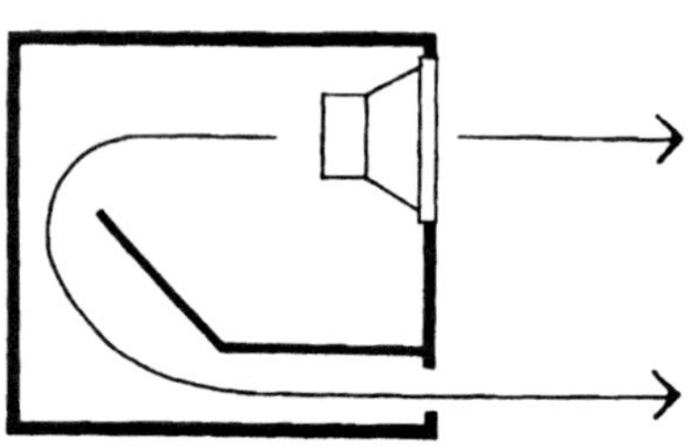

Fig. 28 Bass reflex cabinet.

Closed and open-back cabinets are the most common and most suited for guitar amp speakers. The size of these cabinets depends on the size of the speakers used. Below are common volumes of cabinets.

8" speaker = .5 to 1.5 cubic feet
10" " = 1 to 3 " "
12" " = 2 to 5 " "
15" " = 5 to 10 " "

Bass response can be extended if larger cabinets are used, but huge cabinets are unwieldy. Besides, provisions for very low tones are not necessary when guitars (not basses) are used. There are some very precise theories for devising the effect of cabinet size and shape has upon a sound response. For further information, consult the writings of Thiele, Novak, Small, Newman, and Keele.

SPEAKER MOUNTING

The way a speaker is mounted directly affects its overall performance. There are three ways speakers are mounted: Method #1 offers the most secure mounting without any restriction to the sound; method #2 is less secure but easier to accomplish; method#3 is secure and the easiest way to mount, but sound waves can be disturbed as they pass through the hole in the mounting board.

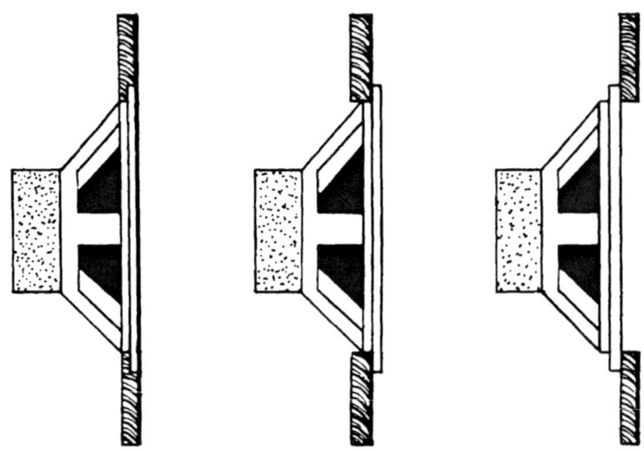

Fig. 29 Mounting a speaker.

Speaker Specifications

Musicians often talk about the "specs" of an amp or guitar's pickups, but they rarely mention the specs of speakers. This, in part, could be due to the fact that sales material for speakers is often lacking in important spec information. People who work with speakers can often benifit from knowing pertinent speaker specification concerning efficiency, frequency response, impedance, directionality of high and low tones, and power handling capability.

Speaker Efficiency

Some speakers may need more watts than others to produce the same sound level. Speaker efficiency is determined by the loudness and set input of a speaker at a given distance. For example, if you stand one meter (approximately 39 inches) away from the front of a speaker, it will produce 96 decibels of sound if it receives 1 watt of power from an amp. Most guitar speakers today generally have an efficiency of 96 to 106 dB of sound output for 1 watt at 1 meter. Twelve decibels may seem like very little difference, but there is actually a great difference. It can be false economy to buy a $100 (96 dB) speaker instead of a $200 (106 dB) speaker because the $100 speaker may require a more powerful, and therefore, more expensive amp to match the sound output of the more expensive speaker (which can be used with a less powerful and less expensive amp). In the long run, a less efficient speaker can end up costing more.

Frequency Response

A perfect frequency response would be a faithful reproduction of the frequencies and loudness of tones. For example, if a guitarist plays a series of notes of varying degrees of loudness, and the speaker identically reproduces these tones, the speaker is said to have a "flat" frequency response. Speakers have an inherent tendency to sound louder at some frequencies than others. This behavior has to be engineered out to achieve a flat response. Since this takes time and money, inexpensive speakers nearly always have an erratic frequency response. Uneven frequency response, even if it is subtle, is annoying and fatiguing to an audience. Summation: When you select a speaker, look for one with a flat response (i.e., not more than + or - 3dB over its range) combined with speaker efficiency (i.e., at least 100 dB with one watt at one meter).

Power Handling

Speakers are rated as to how many watts of amp power they can take. If you have a 45 watt tube amp and a 50 watt rated speaker, you may think you are safe — you are not. This is because wattage ratings are based on watts in the form of sine waves, which are much smoother than music waves. Because music waves greatly vary, depending on the type of music, sine waves are the most reliable gauge for comparing one speaker to another, The initial signal from a rock musician's guitar may start out as a sine wave, but it becomes distorted when processed through an overdriven tube amp. The output of a 45 watt tube amp can reach peaks of 65 watt (or more) square waves. Since a square wave is an abrupt change and not smooth like a sine wave, it is much more stressful to a speaker. A square wave attemps to make a speaker cone move back and forth instantaneously, not smoothly. Sixty-five watts of square wave can be as hard for a speaker to handle as 300 watts of sine wave. Surprising, isn't it? A 45 watt tube amp used for heavy rock music may require a speaker system capable of handling 300 watts or more to achieve speaker safety and non-distorted sound. However, many rockers do want distortion. An overdriven 100 watt Marshall amp plugged into eight 12", 25 watt Celestion speakers can very adequately overload and cause distorted sound. Many people may like the sound, but don't forget that overdriving speakers will shorten their life.

Even though a speaker may handle high wattages, this doesn't mean the speaker will turn those expensive amp watts into high volume. Check efficiency ratings. Cheap, high power speakers waste wattage as heat, instead of creating sound.

Impedance

Impedance is the resistance to a flow of alternating (not direct) current in an electrical circuit. The amount of resistance to a flow is measured in ohms. If electricity flows too easily or not easily enough from an amp through a speaker, it can cause a tube amp to burn up. Therefore, matching the impedance (ohms) between an amp and its speaker is essential to optimum amplifier performance and longevity. For example, an amp which is designed to use a 16 ohm speaker could burn out if connected to a 4 ohm speaker because the current flows through much too quickly, and it overdrives the amps.

Speakers have impedance ratings such as 4, 8, or 16 ohms, but in actuality, speakers do not have just one impedance figure because impedance changes with frequency (represented by Hz). As frequencies rise, so does impedance. If a speaker has an 8 ohm impedance rating, this number merely represents an approximate average impedance for that speaker. An 8 ohm impedance speaker will only have 8 ohms impedance at one frequency, and it could be 2 ohms at a very low frequency. Impedance measurements are usually made somewhere between 100 and 1000 Hz, 400 Hz is common. Two different manufacturers may each sell 8 ohm speakers, but one speaker may be 8 ohms at 100 Hz, and the other may be 8 ohms at 1000 Hz. The former would be a lower impedance speaker than the latter.

If you use a speaker for lower notes than it was designed for, you could blow out your amp. For example, if you play a bass guitar through a lead guitar amp with guitar speakers, the low notes of the bass could reduce speaker impedance so low that the amplifier could be damaged.

SPEAKER UPGRADING

Can better speakers make an amp sound better and louder? The answer is a difficult yes, but the word "better" needs clarification. It is very possible to buy expensive name-brand speakers and have them totally unsuited for certain amps. To find out how speakers perform, connect them to your amp and listen to the sound. Putting a $150 speaker in an old $50 department store solid-state amp is highly questionable; however, putting a super speaker in an old Fender amp could be very rewarding.

Lastly, speaker upgrading can be somewhat deceptive. If a cheap speaker puts all its power into easier to produce mid-range and trebles, it could sound more loud than a quality replacement speaker with a full, solid bass.

Adding Speakers

Hooking additional speakers to an amp will increase the safety margin of speakers used; more speakers can share the wattage. Whether or not stacks of speakers are more beneficial depends on several factors: Are the added speakers efficient? Are they stacked and aimed at the audience? Are the speaker's response curves appropriate for the music being played? Adding speakers can increase sound levels, but it needs to be done in a calculated, not a haphazard way.

When adding speakers, the speaker wiring determines speaker impedance. If speakers are wired in series, you would add the impedance rating of each speaker to determine the total impedance of those speakers. For example, two 8 ohm speakers wired in a series equal 16 ohms total impedance. If speakers are wired in parallel, their impedance is halved (i.e., two 8 ohm speakers wired in parallel will have a total impedance of 4 ohms). This is very important because if you have an amp which specifies use of an 8 ohm speaker, and you use two 8 ohm speakers

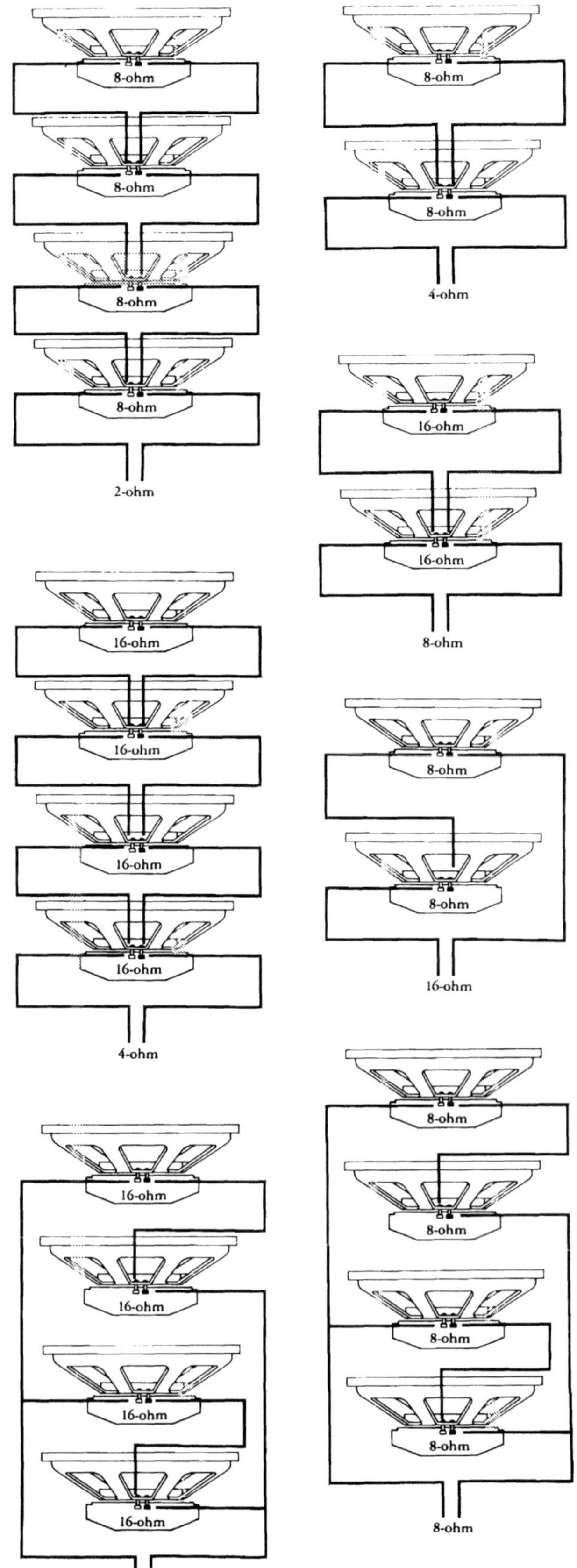

Drawing courtesy of JBL.

Fig. 30 Wiring of multiple speakers.

wired in parallel, you will only have a speaker load of 4 ohms connected to the amp. Low impedances of this level can endanger an amp. Therefore, if you want to use more than one 8 ohm speaker with an amp that calls for an 8 ohm speaker load, you will need to use four 8 ohm speakers; two wired in parallel, and two wired in series. This would result in a total speaker load of 8 ohms.

Speaker Placement

The placement of speakers can boost or reduce volume. A great deal of money can be spent on a high power amp and then wasted by ineffective speaker placement.

For every doubling of distance in an auditorium, the wattage must be increased four times. To adequately project the sound, there are two basic paths to follow: 1) if you are using only a few speakers, tip them back so that the sound bounces off the ceiling instead of becoming absorbed by the audience in the front row; 2) if you are using several speakers, stack the speaker cabinets — you can increase volume 3 dB if you stack the speaker cabinets, rather than place them next to each other.

Also realize that bass tones disperse widely and high notes tend to travel in straight lines. This can result in music sounding bassy near the walls of a room and treble heavy to those in front of the stage. Checking the sound with different speaker positions is helpful.

Sound Balance

Traditionally, the guitar approach to sound is simplistic, and there has been little concern with hi-fidelity. Whereas home sound systems use separate bass, mid-range, and treble speakers, guitar amps usually use only one size speaker for all frequencies.

When one speaker must reproduce the entire frequency by itself, there is a strong likelihood that some distortion will occur. Strong bass notes will conflict with high notes. The result is the Doppler effect — some notes are slightly raised in pitch and some are slightly lowered. By using separate speakers for separate parts of the frequency spectrum, this problem can be avoided.

Speaker Companies

JBL

James B. Lansing's first speaker endeavor was Lansing Manufacturing Company. Later he was vice-president of Altec-Lansing, and after that he formed James B. Lansing Sound, known as JBL.

The first speakers made by JBL were field coil units. When Lansing progressed to permanent magnet speakers, many innovations were developed. JBL was probably the first to make speakers with 4" voice coils. Large voice coils allow a speaker to handle more watts, and the 4" coil JBL-D101 of 1945 was well suited to the high power demands of musical instrument applications. These "D" series speakers were first used in Fender amps in 1963, just two years before CBS bought Fender. The standard D speakers were painted gray, but those supplied to Fender were painted orange. This is how you can tell them apart. The "K" series was introduced in the 60s. The D and K speakers look alike, but there are some internal differences.

When the cost of alnico became prohibitive, JBL began to design speakers that embodied the qualities of the D and K series; the result was the E series which used ceramic magnets. JBL drawings show conventional magnet speaker design and the "symmetrical field geometry" of JBL which intensifies the flux field. SFG allows a ceramic speaker to operate similar to an alnico one.

Celestion-Rola

The Rola Company was originally a U.S. company, but it moved to the U.K. in about 1924. It was in the U.K. that Rola and Celestion became one company, and it was with Vox and Marshall amps that Celestion-Rola G12-M speakers became famous. These blue painted steel frame speakers contribute to the characteristic sound of English amps. The G12-M speakers are not heavy-duty, high wattage speakers, and distortion occurs when given a strong output.

Electro-voice

Electro-voice (EVM) speakers are the second most popular speaker, and perhaps the heaviest duty speaker made. Electro-voice speakers rated at 200 RMS sine wave watts are capable of withstanding short duration peaks of up to 800 watts; for this reason, companies like Boogie and Music Man offer them in their amps.

Gauss

This company has recently entered the premium speaker market. A unique feature in their speakers in the use of two spiders to stabilize the voice coil.

CHAPTER 4

USING AN AMP

What Size Amp

The most important aspect of an amp is its output, not its physical size. The higher the amp's wattage, the greater the volume. Generally, 1 to 20 watt amps are used for practice/rehersal; 20 to 50 watt amps for small clubs/studio use; 50 to 100 watts for small clubs; and over 100 watts for concerts. A general guideline is to use 100 watts per 1000 people.

The term watts can be nebulous because there are different kinds of watt ratings. Power watts, peak music watts or RMS watts. Power watts measure the gross power consumption utilized by the amp; it does not measure the power the amp emits. A tube amp that has a power consumption of 40 watts often sends only 5 watts to the speaker.

Peak music watts rate the power an amp can emit in a split second musical accent. RMS (root means square), a mathematical computation of a sine wave, is the best specification to use. It measures the actual output wattage that an amp can continually emit. However, using RMS to equate tube and solid-state is not always valid because tube amps can be overdriven past their ratings (e.g., a 100 RMS watt tube amp can be pushed to 120 watts). Solid-state amps cannot be readily pushed for extra output because distortion would be too great, and amp failure would result (see chart). When comparing the tube amp Vox AC-30 (with 30 RMS watts watts) to a solid-state amp (with 100 watts), the AC-30 can put out a comparable amount of sound.

The desire for high amp output sound levels is frustrating because doubling the apparent sound of an amp means at least a 4X increase of wattage must be achieved. A 200 watt amp may seem more attractive than a 130 watt amp, but the larger amp will not sound much louder at all — but it is likely to cost substantially more.

Many guitarists are beginning to select several smaller amps chosen for tone and cost, and then miking them when higher sound levels are desired, rather than investing all their money in one large amp.

Choice of an Amp

Early amps were not designed to be used beyond the onset of clipping, or (in guitarist's terms) before the amp was overdriven into distortion. This was because it was assumed that musicians wanted clear sound reproduction, and that distortion was to be avoided.

Guitarists such as Hendrix recognized the tonal possibilities through overstepping the boundaries of clear sound reproduction. Sustain could be increased by overdriving tubes, and feedback generated notes could be produced either acoustically or electronically.

There is a great deal of individual preferences when it comes to tube amps; and there are several basic configurations, each of which contribute to sound.

1. Class: Class A single stage amps vs Class AB push-pull dual-stage amps. Some people prefer the sound of a Class A amp, such as a Vox A30 or a Fender Champ, while others prefer a Class AB amp, such as a Fender Twin. Class A amps give more sustain than a Class AB push-pull circuit. This is because the Class A amp always has full current. When you hit a chord on your guitar, the Class A amp doesn't have to build up current because it's already there.
A 30 watt Class A amp could sound as loud as a 60 watt Class AB amp.

2. Tube Design: When it comes to tube preference, it needs to be noted that most people only disagree about which power tubes are best, not which pre-amp tube is best. However, there are still many types of power tubes to choose from.

Some feel that 6V6 Tubes have a smoother sound than the more powerful 6L6 tubes and choose tone over output power.

3. Tube Type: A 6L6 made in one country can be substantially different from one made in another country. Also, even in one country, each tube company uses its own configurations. For these reasons, the brand of tube and country of origin can make certain tubes sound warmer and sustain longer. Since manufacturing procedures are constantly changing, it is impossible to write a relevant guide to tube origin or brand performance.

If two different name-brand amps both use 6L6GC tubes of Sylvania origin, it does not mean they will sound the same. This is because there are almost an infinite number of ways of putting together an amplifier. Sound is very subjective, and this is why there are so many companies building amps.

Playing an Amp

An amp's controls, e.g., the volume, tone, and master volume, greatly influence the sound it produces. It is important for a musician to learn how these controls effect the sound of an amp. Giving an amp a strong, full-volume input from a guitar while having the amp set at low volume, will overdrive the input of the amp and produce a low volume, heavily distorted output. Giving an amp a low volume guitar input with the amp set on full volume, should produce a loud, clear sound.

Many tonal variations are possible with an amp. The addition of presence and mid-range tone controls, input-gain, and master volume controls, all contribute to the shaping of sound. Basically, overloading any tube in an amp will result in distortion, and an increase in sustain. Heavy sustain is generally created in the pre-amp section of an amp by feeding (cascading) the output of one pre-amp tube into the input of another pre-amp tube. Amps with two or more pre-amp tubes are the norm for tube amps; some amps even have four or more cascading pre-amp tubes. If an amp has volume controls between these tubes, a great deal of tonal flexibility is possible. Amps with more than one gain control are built around this premise.

Experimentation with different settings is the teacher. Be creative and explore all sounds possible.

Influences to an Amp's Output

INPUT INFLUENCES

Input voltage plays a crucial role in determining the output wattage of an amplifier. If a 100 watt amp is designed to work with a 100 millivolt guitar input voltage, but receives an input of only 50 millivolts, that amp will not be able to put out a full 100 watts. The use of high output guitar pickups and/or pre-amp boosters can raise a guitar's output voltage to a sufficient level. The reason for high guitar output is to achieve the overdriven, compressed sound which is characteristic of tube amps. Note: Caution needs to be exercised because excessively high guitar output can damage an amp. Most guitars have an output of less than one volt and most amps can safely take an input of 1 to 3 volts (check with the manufacturer). It is the pre-amp boosters that can cause problems, because some of these put out 3 or more volts.

OUTPUT INFLUENCES

High speaker impedance constrains an amp's output wattage and limits volume. However, if the speaker impedance is too low, this can also cause problems. For example, if a 100 watt amp designed for a 16 ohm speaker is plugged into a 4 ohm speaker, it may put out 140 watts, but the tubes and transformers will overheat.

When putting together your own system, make sure you have the correct amplifier and speaker impedances. Speaker impedance is not D.C. resistance; therefore, if you use an ohmmeter to measure the speaker resistance, it will be approximately 3/4 of the impedance rating. For example, an 8 ohm speaker will measure approximately 6 ohms.

Areas of Amp Problems

An amplifier problem can start as a barely noticed fault and eventually progress to complete amp failure. Putting off repairing a minor problem can be a major mistake.

The causes for problems in tube amps are heat and vibration. Normal heat wears out tubes, and excessive heat can burn out transformers, resistors, and diodes as well as tubes. The main causes of excessive heat are short circuits in the amplifier, mismatched speaker impedances, and playing at very high volume for long periods of time. Below are ten of the most common amp problems.

1. Unplugged amp. Yes, the most common cause of amp failure is an unplugged amp.

2. Fuse. A blown fuse will turn off an amp. A fuse may blow for no reason, or for non-harmful high voltage music peak. If a fuse is replaced several times and immediately blows, there is likely to be a serious problem within the amplifier.

3. Tubes. An old tube will have low power and will distort easily. Old or new tubes with loose internal structures will be microphonic, causing annoying squeals and ringing, especially at high volume levels. A tube that has totally broken down inside will often cause a short circuit and blow a fuse.

When replacing tubes, use a new tube of the same designation as the old one, i.e., replace a 6L6GC with a 6L6GC. It doesn't matter if the tube is made by another company; however, there are two exceptions when replacing tubes: 1) Always replace a 7025 with a 7025. You can replace a 12AX7 with a 7025, since the latter is a higher grade of the former; but do not replace a 7025 with a 12AX7. 2) You can replace a 6L6GC with a STR415, since the latter is a higher grade than the former.

4. Power loss. Power loss is the most common malady of a tube amp. If an amp sounded great at the shop, but very bad while playing at a gig, this is likely due to power loss. A club's air-conditioning or heater, lights, etc., could drain the house current from 120 volts to 100 volts. This would result in a 50% power loss (i.e., a 100 watt amp would only put out 50 watts).

It is advisable for tube amp owners to have their own variac that can be set to obtain a consistent 120 volts. These sell as low as $80.00 and all you have to do is set it at 120 volts and push the button.

Bad tubes, a too high speaker impedance, or a too low input to the amp will result in power loss.

Often, the buyer of an old tube amp takes it to a shop for an overhaul/check out, which includes new tubes. The owner of the amp then plays through it and complains that the good old amp sound is gone and blames the shop repairman for faulty work. The reality is that the amp owner liked the sound of old worn out tubes that were half disintegrated. New tubes will be louder and sound clearer, i.e., distort less, than old tubes.

5. Hum. A club with fluorescent lights. light dimmers, and nearby electric motors will often cause an amp to pick up hum. Playing non-humbucking pickup guitars will also cause hum. Sometimes the most practical way to reduce hum is to move the amp or guitar player to a different place on the stage. When an optimum position is found, recheck the polarity switch on back of the amp to see if throwing the switch (to reverse the ground) will help reduce hum. Very loud hum can also be the result of a broken ground connection.

6. Noisy controls. Volume and tone controls, which crackle when they are turned, are likely to be filled with dust. Controls can be cleaned by spraying Electronic Parts Cleaner into the control potentiometer as they are turned back and forth. Controls which still make noise should be replaced.

7. Transformers. Incorrect power levels can result in fried transformers. An overheated transformer will look and smell burnt. The failure of a transformer is sometimes accompanied with a cloud of smoke. Another transformer problem results from loose mounting bolts which allow the transformer to bounce around and damage small components.

8. Filter capacitors. The large paper capacitors used in the power supply sections of older amps can cause amp malfunctions as well as loss of safe amp grounding If leaking fluid is detected from filter capacitors, replace them. If an amp is not used for a long period of time, paper capacitors may start to leak.

9. Broken leads. Vibration and rough handling can result in broken connections between amp components. A loose solder joint can seem alright upon a glance, but a crack in a printed circuit board can be so small that can't be seen.

10. Reverb. Non-functioning reverb units usually have a broken lead in the input or output transducers. The transducers or springs don't often break.

CHAPTER 5

AMP COMPANIES

DOBRO-NATIONAL-VALCO – RICKENBACKER

If anyone studies the first electric guitars and amps made (the ones debuting circa 1931), one can be perplexed about the offerings of different brands of instruments: Dobro, National, Valco, and Rickenbacker. Old catalogs reveal that these companies each had a model corresponding to that of the other. At first it seems that one simply copied the other.

National was formed around 1924 in Los Angeles by John Dopera, a Mr. Kilmar, and a Mr. Beecham. Shortly thereafter, John left National with his brothers, and formed the Dobro Company.

Paul Barth, the son of the foreman of the National Company woodshop, got together with George Beauchamp of the Rickenbacher Company, and together they made an experimental electric guitar in 1929. This guitar used a National style guitar neck made by Paul's father. The pickup was made by Beauchamp. This famous guitar is often shown being held by Adolf Rickenbacher, and the result was that Rickenbacker got credit for making the first electric guitar. This mistake is easy to understand, especially since National was not strongly interested in manufacturing electrics; but Rickenbacker was. Interestingly, the first patented electric guitars of Dobro and Rickenbacker were in almost dead heat competition.

The amps used for these instruments and other early electrics were AM radio circuits with the receiver section removed so that only the audio amplifier and speaker remained. In fact, many of the first electric guitar amps were no more than homemade adaptations of radios.

Rickenbacker unveiled their electric guitar and amp in 1931. A short time later, Dobro produced electric guitars and five tube, ten watt amps. Both these guitar/amp sets sold for around $100. By 1982 standards that would be approximately $1,000. This was a high price, but these instrument/amp offerings were the highest technology of the day. A year later, in 1932, Dobro, National, and Rickenbacker offered guitar amps. Most of these amps' chassis were made by a Los Angeles radio supply company.

The early Dobro and National amps and instruments, along with those of Rickenbacker, did achieve nation-wide recognition. Although the early, primitive amps had low output and had to rely on cumbersome field coil speakers, they did show the world that the electric guitar and amp had been born.

Fig. 31 Rickenbacher amp of the early 1930's.

Fig. 32 Rickenbacher nameplate. The name was changed to Rickenbacker during World War II.

Fig. 33 Rickenbacker amp of the late 1950's.

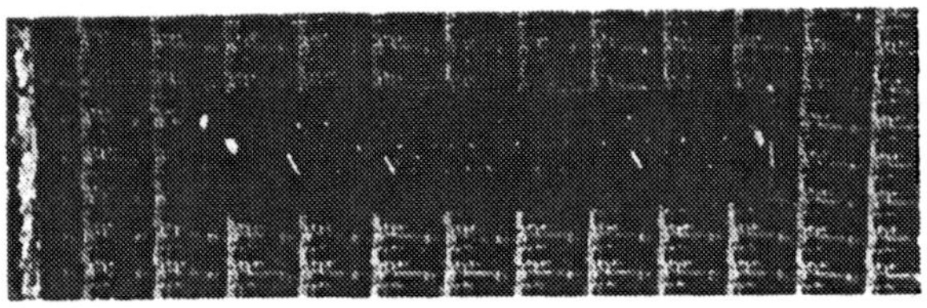

Fig. 34 Rickenbacker nameplate.

During the thirties, Rickenbacker guitars and amps were popular; during the forties and fifties, other gained the spotlight. In the 60's, Rickenbacker gained new popularity with their guitars and tube amps. Rickenbacker does not currently make tube amps.

Around 1934 Dobro and National had moved to Chicago, and the rights to the Dobro name were licensed to Regal. The National Company floundered in 1939, and after World War II Louie Dopera, Al Frost, and Victor Smith formed VALCO (Victor, Al, Louie, and Company). One of Valco's line names, interestingly enough, was National. National made guitars and amps for Montgomery Wards, Sears, Spiegel, and others.

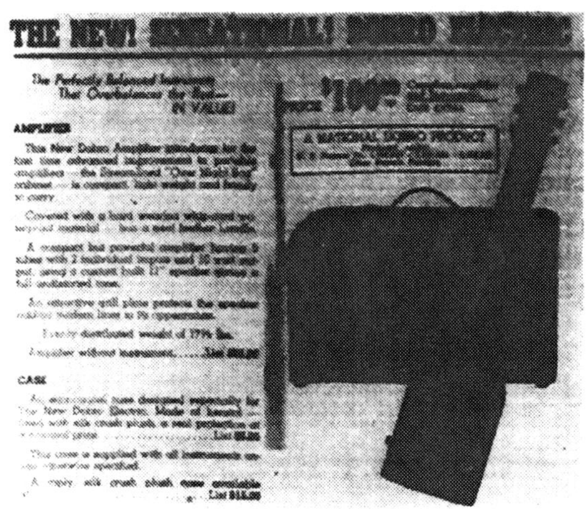

Fig. 35 Dobro amp of 1932.

Fig. 36 National amp.

Valco is a confusing company to chart, because they sold amps under many different names as model lines for Valco. In addition, Valco was contracted to make amps for other companies: Oahu, Barnum, Sonex, Sound City, and possibly even Gibson. The amps made for chain stores usually had no brand name on the front and only a small label in back with the imprint, "Made by Valco". One exception was a series made for Montgomery Wards, which carried the Magnatone nameplate.

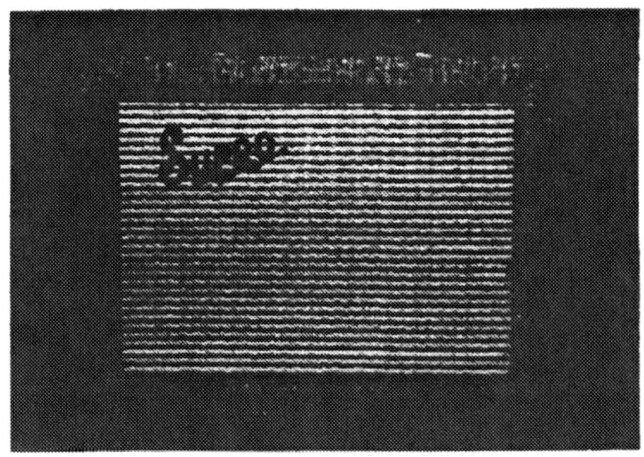

Fig. 37 Supro amp.

The Oahu and Supro model amps were some of the first amps with tremolo and reverb. Early Supro amps had a Supro nameplate in the center of the grill, the name Valco below, and the shield design of National on the side. The Supro line was quite extensive, with model names such as "Thunderbolt". The size of these amps ranged from 5 to 30 watts.

All the Valco amps were designed by Victor Smith, and most were based on Fender designs. Whereas Dobro had limited success with their amps, Valco enjoyed commercial success. Valco continued to make amps for many companies for many years.

In the 60's a series of amps called the Glenwood Professional Series, with National nameplates, were distributed by Fred Gretsch. These amps were companions of the modernistic National Glenwood guitars. The National Glenwood tube amp line-up consisted of nine models, from 5 watts to 70 watts. All cabinets were covered in black, white, and silver Tolex/Lumite, with Saran fabric grill cloth. All amps had Jensen speakers. Today both Valco/National amps and electric guitars are prized by many. The old, easily overdriven tube sound of the small amps is miked and run through a P.A. for high volume levels. Two of the most desired National/ Valco made amps are the Supro and the Oahu. The Oahu name clues us to the fact that their early amps were made primarily for use with Hawaiian guitars.

OAHU

Oahu amps came out of the Oahu Publishing Company in Cleveland, Ohio. The earliest Oahu amp known, model 230K, is from the 1930s. (This old amp is quite primitive in appearance). It is believed it could have been made by Lyon and Healy. During the 40s and 50s Oahu had Valco make amps for them. The Oahu line-up consisted of many different models, most of which were in the range of 5 to 25 watts. The ones made by Valco are the most desired; ones not made by Valco are considered quite dull. But all Oahu amps of the 30s and 40s lacked full bass or high treble tone due to the small, inexpensive speakers used. When connected to a good 10 or 12 inch speaker, these amps can be overdriven to produce a full, warm vintage tube sound.

Fig. 38 Oahu amp.

FENDER

Since Leo Fender was working in the amplifier field long before he founded the Fender company, these years are essential to amp history.

Leo was a bookkeeper in the 20s, and he began a hobby of building and repairing radios. He built radio gear for friends, and in 1929 he put together a ham station. A few years later, when he was doing civil service accounting work, Leo rented out the three P.A. systems that he built for use at ball games, horse races, and music gatherings. He began to spend more and more time with

Fig. 39 Leo Fender.

electronic work, and eventually he became a radio repairman. He ran a Radio and Television Supply in Santa Ana, California (about 10 miles southeast of Los Angeles) on Oak and First. One day, in the early forties, an electronic parts rep, Doc Kaufman, came into the store and began to talk to Leo about electric guitars and guitar amps. Together, Leo and Doc started a new company — K&F Manufacturing Company. They built an electric lap steel and small amp in 1944, and decided to manufacture electric guitars and amps. The first commercial K&F amps came out in 1945; they had a gray, wrinkled finish put over a pine cabinet. Over the next couple of years, approximately a thousand of these 4 watt wonders with gray finish were made. About 12 were made with linen tweed.

Fig. 40 Doc Kauffman and first K&F model amp.

In 1946, K&F was making only one basic amp and lap steel guitar. Leo wanted to expand the company and designed two more amps: A model #2 and a model #3.

Fig. 41 Front and back of later K&F amp.

Doc was hesitant to expand K&F so Leo went off on his own to form Fender Ėlectric Instrument Company. In early 1947 the first Fender Company amp, model 26, was made; it was the first Fender amp to be produced in quantities. (There were a few on/off models, but most have no nameplates or model number.) The Model 26 amps were made with red/orange, yellow or blue colored cloth to cover the speaker. True grill cloth wasn't yet made. For speaker protection, these amps had two thin and one wide strip of chromed metal positioned vertically in front of the speaker cloth. The Model 26 amps have 6" speakers, and are forerunners of Champ amps.

In late 1947 the first fully designated Fender amp line-up was advertised. At this time, the amps were the model 1 Champion 600, the model 2 Princeton, and the model 3 Deluxe. It was in the first of 1948 that Fender introduced its first dual speaker amp: the Model 4 Dual Professional. This amp was sold for use with steel guitars, and it had a slanted front housing with two 10" speakers. In 1950 the amp's name was changed to the Super Amp and retained the angled cabinet. A single speaker version of this amp without the angled front was sold as the Pro Amp.

The first major change in Fender amps was the beginning of a switch from metal cased military tubes to glass subminiature tubes. The 6N6 and other strange old tubes became less common. Remember, this was post World War II, and everything was made for military specifications. Industry had not yet geared up to make all tubes with consumer specifications.

Fender began with the most rudimentary amp circuit consisting of a minimum of parts: a preamp tube, power tube, and rectifier tube. The first truly commercial Fender amp was the three tube Fender Champion 600 amplifier. Interestingly enough, for over thirty years this amp was manufactured and remained popular. Early Fender amps had a basic "no frills" circuit based on components available at the time. The tubes were military type units, left over from the war. The controls of these amps often consisted of an on/off switch built into a volume control.

Fender schematics dated in late 1953 and early 1954 show a switch from the old military metal case 6SJ7 tube to a modern twin stage glass cased 12AX7. These dates for this change-over are questionable. It is possible that the factory change occured as early as 1950, but that the redrawn schematics did not enter service books until 1954. (Note: although many old Fender amps now have glass tubes, most pre 1952 models came with metal encased military-type tubes.)
The use of 12AX7 tubes increased the boosting power and allowed for the easy inclusion of treble and bass controls in the circuit. Many of the first Fender amps to receive 12AX7 tubes did not immediately receive new tone controls. It wasn't until 1954-57 that bass and treble controls were added to most models. It was in 1955 that a presence control appeared, and it was first used on the Pro-Amp, the 3/10 Bandmaster, and 4/10 Super. The tone controls of Fender amps (located between the first two stages of gain) are passive. They don't add, they can only subtract. To achieve a flat response, all tone controls (not volume controls) should be set to full on; that is #12 on old amps and #10 on newer models. Also, in the mid fifties, Fender amps' wattage ratings referred to the RMS watt output, not the wattage consumed by the amp. The very early Champs have 25 watts inscribed on the back of the chassis or schematic, but this was its power consumption; actual output was approximately 4 watts.

The actual output wattage of Fender amps, especially the 1950 15 watt "Super" amp, was too much for many speakers of that time (usually Jensen speakers were used). For this reason, the Super (and later the Twin) needed two speakers to share the load. Around 1958, Fender began to also use Oxford, CTS, JBL, Utah, and Trusonic.

It was in the mid fifties that a rapid increase of power occured. The change from 6SJ7 to 12AX7 pre-amp tubes was the first step; now amps began to receive additional tubes. In 1952 Fender made its first amps with multiple input-stage tubes. These were the Twin (three for each channel), the Bandmaster (three for each channel), and the Deluxe (two for each channel). The first separate treble and bass tone controls were used in 1953; the first amp to get them was the Twin. 1954 was the year that the first presence control was introduced. First applied to the Twin, it was soon added to other models. In 1955 some Fender amps began to receive vibrato circuits through the use of an added 12AX7, which functioned as an ocillator. Almost all Fender amps became available with vibrato, even the lowly Champ (in 1964) became available as the Vibro-Champ.

Note that Fender has used the terms vibrato and tremolo indiscriminately. Vibrato is actual pitch change, whereas tremolo is fluctuations in volume. No Fender amp has vibrato, only tremolo. The first Fender amp with tremolo was the 1955 Tremolux model 5G9-A.

In 1956, again with the Twin, a new control debuted for Fender; this was the mid-range control. In later years, it has been applied to other models.

Also during the fifties, there were refinements in the power supply, and filters were added to smooth the current flow. The major changes in Fender amps, though, were to come in the 60s, when Fender began a change-over from tube to solid state rectifiers.

Some think a change from tube to solid state rectifiers made a big difference, but this is because there were also changes made in other parts of the circuit at the same time. The first year a Twin had a solid state rectifier was in 1960, yet the 1963-66 models are the "good ol' Twins" that many preferred.

Fender amps (Vibroverb and Twin) received reverb circuits in 1963. A 12AT7 was used to boost the signal into the reverb unit, and half of a 7025 (12AX7) was used to boost the signal coming out of the reverb unit. 1958 was the year of the first 100 watt Fender amp. The Twin had now graduated to four 6L6GC power output tubes. The year of the great change-over was in 1965 when CBS acquired ownership of Fender. Many people still claim that CBS ruined the quality of Fender instruments and amps, but this doesn not take into consideration the full story. Vince Basse, who used to be a Fender Service Center manager says: "I think it is hokum to say it [the CBS take-over] changed the amps' tonality. The amps did have a tonality change, but it had nothing to do with the CBS take-over because these changes were already in the works prior to CBS's involvement."

One of the most noticed changes was with the Twin. In 1965 the Twin went from 100 watts down to 85 watts. (Although some Twins had this modification, some did not.) The reason for this decrease was that Jensen speakers used at that time could not handle 50 watts each. The speaker distorted under full load (and eventually failed), but some people liked its distorted sound. The power reduction was achieved by putting a 7 watt resistor in the output; if this resistor is removed, these amps will give a 100 watt output. A switch to other speakers, most notably JBL (first used in the mid sixties) allowed power to be increased.

In 1969 Fender amps using dual output tubes received a tube matching circuit. This allowed an amp user to use lower costing, non-matched tubes with no compromise in sound. A chassis mount trimmer pot is used as the control to balance, therefore match, the output of the tubes.

Two, rather humorous additions, to the newer Fender circuits are the "bright switch" and the "boost switch". These controls are often thought of as being too limited in their effect. The bright switch works in this way: In the bright mode the amp is actually in normal, in the non-bright mode the highs are shorted to ground, and this results in a non-bright sound. This is often referred to as a "dull defeat tone control". The boost switch offers only a moderate amount of boost. It is sometimes referred to as the distortion switch, and is built into the volume control as a push-pull function of that knob. Fender amp designers felt that since boost circuit uses the reverb circuit (it overdrives the pre-amp tube), they didn't want to greatly overdrive the tube and severely shorten its life.

The most complex change in Fender circuits would be the Fender 75. This amp has a wide range of the tube sounds. See the section of the 75 for details.

In summation, Fender circuits are really very basic. Tube amps are not necessarily all that exotic, but Fender definetly set the pace and style of amps. Years ago the advice was: "If you build an amp, make sure everything is at least twice as strong as it has to be". Fender amps have certainly stood the test of time.

Basic Fender Amp Circuits

Fender made many models of amps, but it is important to realize that many amps were identical inside; the differences were outside, e.g., number and size of speakers and cabinet design.

For example, the chassis in the Twin is basically the same as that of the Showman Reverb, Vibrosonic Reverb, Super Six Reverb, and Quad Reverb.

The Super, Pro, Bandmaster, Concert, and Vibrosonic are all of the same chassis family. Over the years, when one was changed, the others were also, with the exception of the Super (see that heading). The reverb family of the Super Reverb, Pro Reverb, and Bandmaster Reverb likewise shared a similar chassis.

When comparing Fender amps it is useful to remember that out of 15 amp models, there are only four basic amp configurations, but those four designs have been modified over the years, and sometimes even changed during a single year.

EVOLUTION OF FENDER AMP EXTERIORS

Fender Nameplates

The familiar Fender nameplate on the front of a Fender amp wasn't always familiar; some of the first amps Fender made did not even have a nameplate, e.g., the Model 26. The first Fender amps with names, made after the Model 26, had the model name in script and was silk-screened on the control plate.

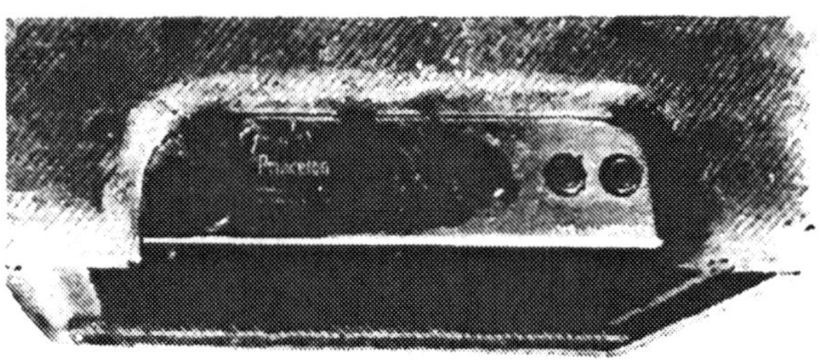

Fig. 42 Control panel of TV front Princeton.

The first Fender nameplates (1948) were made of small strips of brass which had the background silkscreened a very dark brown, and brass block letters spelling Fender. In 1951 a switch was made to the famous Fender script, using the backwards "F" design; but as late as 1959, block design Fender nameplates were used.

These plates were attached to the front center of the amps, just above the grill cloth. Some plates included the name of the amp model (this was primarily the case of top of the line models), but some did not. Also, some included the words "Fullerton, California", but some did not.

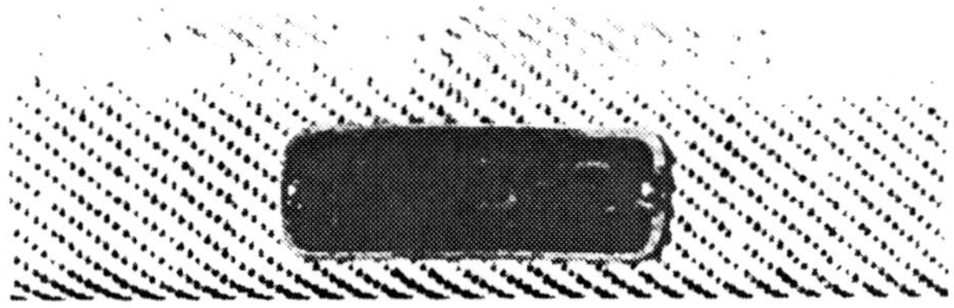

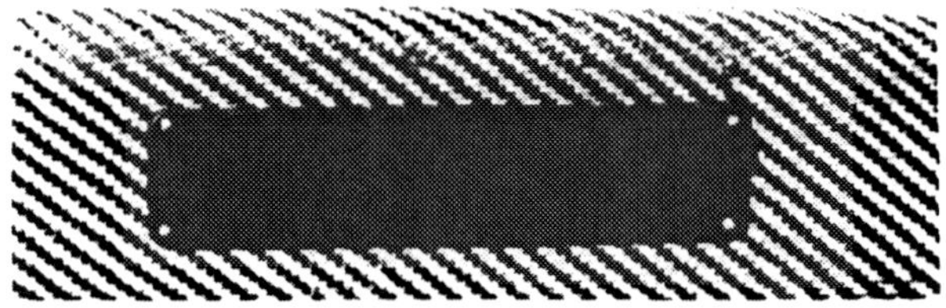

Fig. 43 Block letter nameplate.

Fig. 44 Script letter nameplate.

Fig. 45 Script nameplate with model name.

Around 1960, Fender attached the first die cast "Fender" logo to the grill of an amp. In 1963 this flat logo was changed to a more sculptured stamping which had a black painted inner section. Fender changed the logo in 1964 by deleting the tail of the R. This is the logo that has become standard.

Fig. 46 Flat metal nameplate.

Fig. 47 Sculptured nameplate with tail.

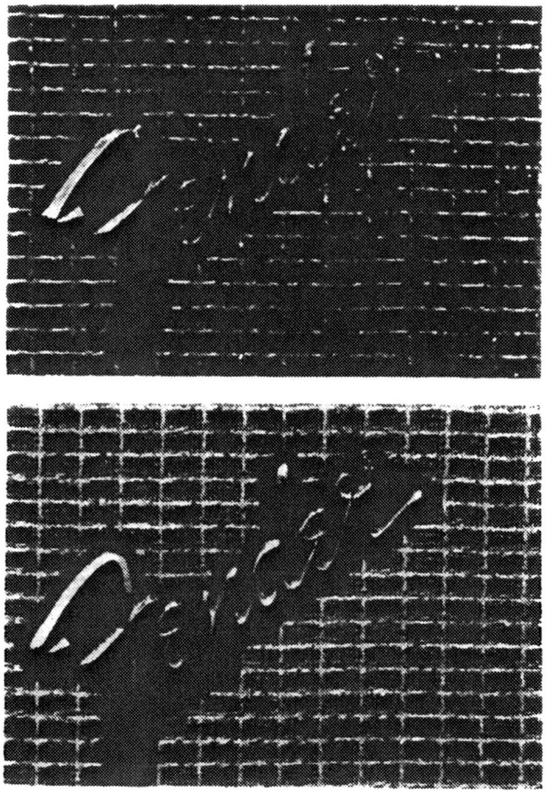

Fig. 48 Sculptured nameplates without tail. Example on top was first used on the last 1960's blackface models and was standard on silver-face amps. Example below was first widely used around 1981 and is used on new blackface amps.

Coverings

Though the earliest Fender amps had plain wood cabinets and early Champ amps (made in 1948 and 1949) had two-tone coverings, Fender "tweed" has become synonymous with vintage Fender amps. This is because of all the older Fender amps, the tweed amps are the most prized.

Fig. 49. Fender Tweeds: #1 Original pure white tweed, #2 Diagonal brown with white dot #3 Diagonal white with light brown, # 4 Diagonal white with dark brown

Shown here are black and white reproductions of Fender tweed. Remember, age and exposure to the elements effect the color. The tweed had the trade name Peroxaline, and was composed of luggage linen treated with nitro-cellulose lacquer. Although fairly durable, this fabric was stiff and difficult to bend around corners.

Fender tweeds, which debuted in 1948, have gone through many changes, which came about because of supply. Also, the tan or pale gold of old tweed started as pure white, but age caused the lacquer to darken. The very first Fender tweed was a pure white (soon aged to golden tan) luggage linen. In 1948 a dot-like white and brown tweed was used. Around 1950 a diagonal white and light brown tweed was used, and shortly thereafter there was a switch to white and dark brown diagonal tweed. This last type is the most common. At least four different types of tweeds are now known, and it is possible that a few other odd patterns may have been used when the standard materials were not available.

The Champion 600 amps were first made around 1948, the same year that the Broadcaster electric guitar came out. The guitar came with a two-tone case made of naugahyde-type fabric of tan and dark chocolate brown. Champion amps were made with similar coverings. In late 1949, when the Broadcaster name was changed to Telecaster, cases and amp coverings changed to "aircraft luggage linen". This covering, referred to as Fender tweed, isn't tweed; but the name is now entrenched by usage. These guitar cases, with tweed fabric of very pale gold and dot stitched bands of dark brown, were supplied by Bulwin Manufacturing of Costa Mesa, California. The very first Fender amps, however, had a covering of uniformly pale tweed. When new, this tweed was nearly white, but age has given this fabric a deep golden color. In a short period of time, the guitar case tweed was used.

In about 1950, this tweed was changed to pale gold with continuous zig-zag patterned strips of brown. The amps also went from square corners to rounded corners; but the corners remained exposed without protective covers.

In early 1960 Fender redesigned many of its amps and some of its models (the Super, Pro, Concert, Bandmaster, and Vibrosonic) with brown Tolex cloth that had a coarse, almost burlap-type grain pattern. The lower amps (Champ, Tremolux, Deluxe, Harvard, and Vibrolux) remained with tweed. In 1961 off-white Tolex was used to cover what Fender termed as its professional model line: the Twin, Showman, Bandmaster,

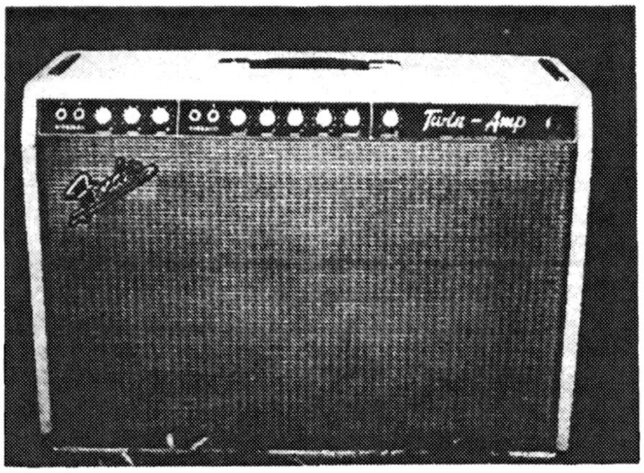

Fig. 50 White Tolex covered Twin amp.

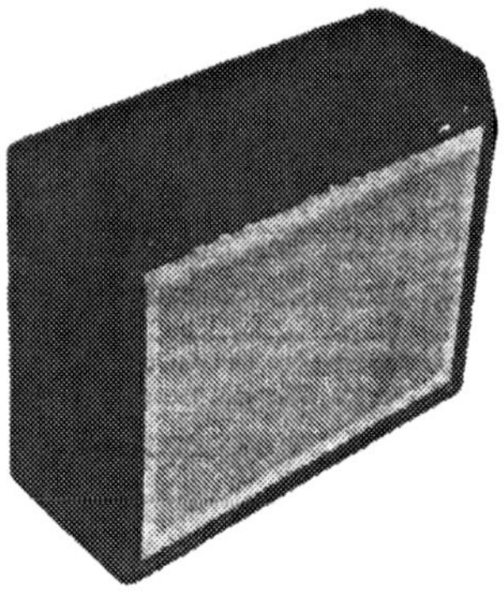

Fig. 51 Brown Tolex covered Vibrolux amp.

Tremolux, and Bassman. This covering is often called white or cream, or light vinyl vinyl. In 1963 Fender introduced black Tolex. In 1963 Fender introduced black Tolex covered amps — the Twin amp Model AA763. In 1964 Fender introduced black Tolex on the newly redesigned Champ, Harvard, Deluxe, and Pro, making 1963 the last year for tweed models. In 1964 a new type of brown Tolex was used on mid-range models, and this covering had a pebbled, leather-like grain. At the end of 1964 the white and brown coverings were discontinued and all amps went to black Tolex.

Many Fender amps have gone through an evolution of several coverings. For instance, the Fender Twin debuted in a whitish tweed (1952), then went to a white and brown dot tweed (1953), and finally to a white with dark brown zig-zag tweed (1954). After this, Twins switched to a variety of Tolex fabrics: cream Tolex (1960), and finally to black Tolex (1963). A brown Twin is said to exist, but this must have been a special order or a recovered model.

Control Panels and Knobs

The earliest Fender amps, like the K&F amps, had the control/jack panel in the back of the amp. The amp chassis was in the bottom of the cabinet, and the controls were in the upper section. When the company integrated the control panel with the amp chassis, only the position of the chassis moved, not the controls. These first control panels were usually left as bare metal, but sometimes they were painted black.

Around 1950, Fender amps had back side control panels painted copper with white lettering. These amps had either black pointer knobs, or later, round knobs with one indent stripe down the knob's center. Both these knobs were standard off-the-shelf items. (Though it is not substantiated, it is possible that some of the Model 26 amps did have the same round knobs with three central indentation stripes similar to the K&F amps). The numbers 1 to 12 were printed on the control panel, and not on the knobs. This numbering method continued until the mid 60s. The control panel was moved to the upper rear portion (circa 1951). The back edge of the top was cut away to reveal the chrome plated control panel/amp chassis, which had black pointer knobs and white lettering.

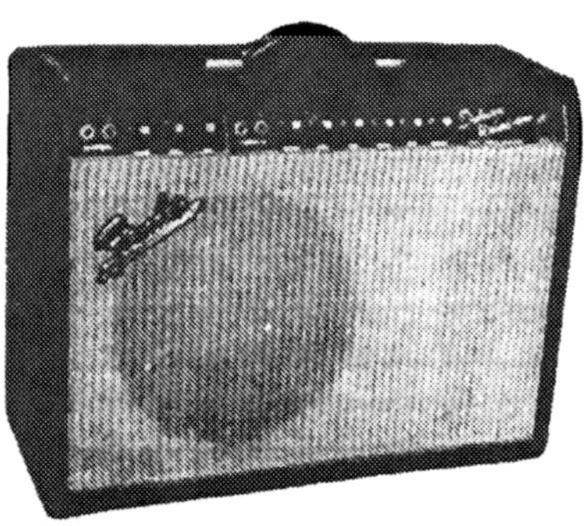

Fig. 52 Black Tolex covered amp, a Deluxe Reverb.

Fig. 53 Copper and White Champion 600 control plate.

Fig. 54 Control plate for tweed Champion amp.

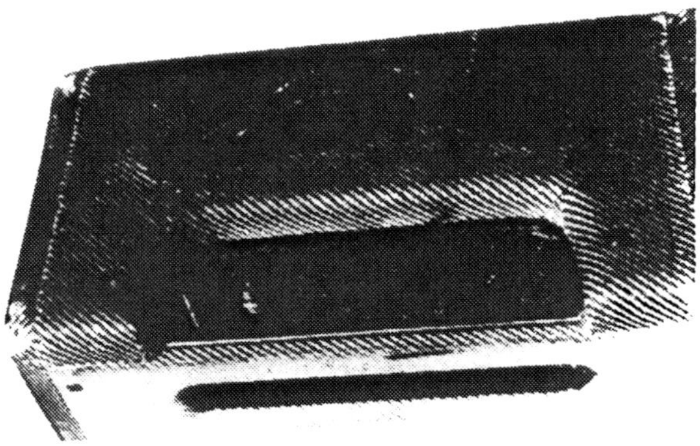

Fig. 55 A top/rear cutaway control panel.

In 1960 Fender began a redesigning of its amp cabinets. The control panels were put on the front and painted dark brown; the lettering was the silver of the bare metal plate. These amps were called brownface. The knobs on these first amps were round/dark brown and later were changed to white plastic.

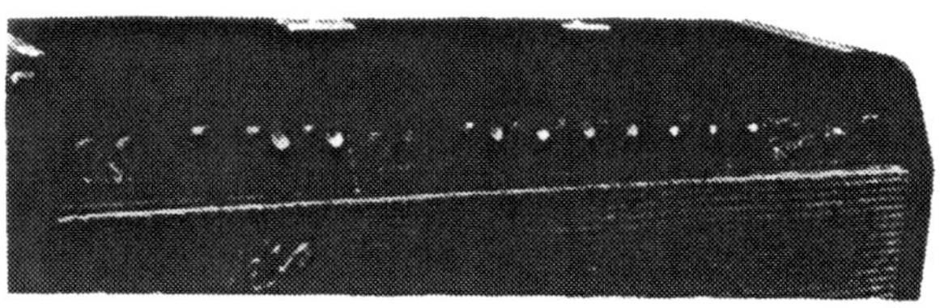

Fig. 56 Blackface control panel.

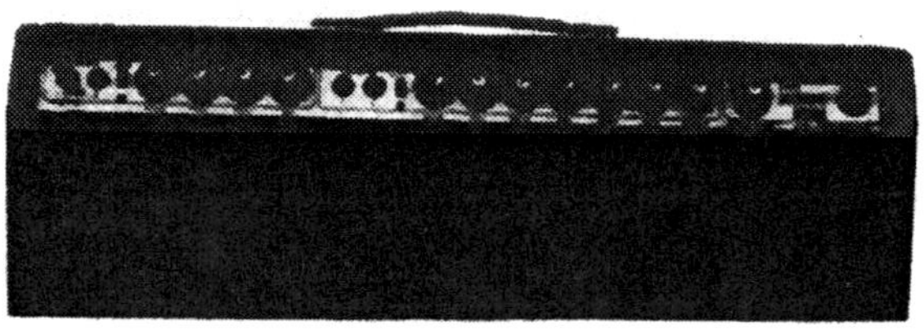

Fig. 57 Silverface (whiteface) control panel.

The term blackface occurred when, in 1963, Fender amps went to a black painted control panel. Although blackface is a term often used to divide "good ol' vintage Fenders" from "that CBS stuff", it is enlightening to find "good ol' blackface amps", with date codes that show they were made as late as 1968 (three years after CBS bought Fender).

When Fender went to blackface, some models did get cream/white round knobs; but the standard knob was a black knob with a skirt lip that had white numbers printed on it. This is the type of knob still used today. A few Fender blackface amps have been found with anodized aluminum knobs and others with skirted white knobs with black numbers; very similar to those which were used on the Jaguar guitar. It is possible that these were installed by past owners of the amps, or they might have been factory installed when the stock of standard knobs was depleted. In 1969 there was a transition from blackface to whiteface, i.e., a silver metal control panel. A few amps were made with the new turquoise lettering printed on a blackface panel.

In 1970 Fender went to an aluminum control panel with torquoise stripes and block lettering. This is the type of panel that was on all Fender amps up until 1979 when Fender reintroduced blackface panels on the "75" amp. The knobs have remained the same. Fender changed its entire amp line to new blackface in 1982. The quickest way to tell a new blackface from an old one is that the newer amps have the Fender script logo without the flourish at the tail of the "r".

Fig. 58 New blackface control panel.

Grill Cloth

The development of a specialized, tough, non-elastic, non-absorptive cloth did not occur until the early 1950s. This was the invention of Saran, a trade name for the first grill cloth.

The first Fender amps used ordinary cloth for speaker protection. Three strips of chrome plated metal were attached in front of the speaker on top of the cloth. Leo said there was no significance to the color of the cloth used on those first wood boxed amps. Models have been seen with scarlet cloth, blue cloth and yellow cloth.

Starting with the Champ in 1947, Fender amps had cloth that was a very dark brown. Most of these amps today have rose colored cloth due to the fact that the color has faded. These amps had no metal strip speaker protection.

Tweed amps were first made with dark brown, jersey fabric grill cloth. In 1952 true Saran grill cloth was introduced; the first examples were solid dark brown. Soon, dark brown grill cloth with threads of yellow tan were used. This was followed by a gold colored cloth with threads of reddish, dark brown. In 1955 sculptured (slightly three dimensional) grill cloth was adopted.

Maroon cloth was used from 1960-63 and from 1962-64 gold cloth was used.

White Tolex covered amps were first made with dark brown cloth with threads of tan. Later, white amps went to the solid tan cloth that is most often seen.

Brown Tolex amps initially had brown cloth with pale threads; later, they received cloth composed of tan with dark threads.

Black Tolex amps (in 1963) had gray colored cloth with clear and black threads. The clear threads gave an impression of silver. Some time later, silver-colored threads replaced the clear. In 1969 the first gray and silver cloth with blue threads was used. A cloth utilizing more blue color became the standard in 1971, but a cloth of silver and black with red threads was used over a period of about one year. This cloth was left over from Fender's solid state Zodiac amp line. Black grill cloth was first used on the Super Twin and Studio bass, but the standard has remained silver and blue.

Related to grill cloth are the silver strips which outlined the grill. This ornamentation was used mainly in 1969.

Model Numbered Amps

#	Model	Year
# 1	Champ	1947
2	Princeton	1947
3	Deluxe	1947
4	Super (1949 as Dual Professional)	1950
5	Pro	1950
6	Bassman	1951
7	Bandmaster	1953
8	Twin (2 prototypes, production in 1953)	1952
9	Tremolux	1955
10	Harvard	1956
11	Vibrolux	1956
12	Concert	1957
13	Vibrosonic	1957
14	Showman	1961
15	Reverb	1961
16	Vibroverb	1962

Non-Model Numbered Amps Order of Debut

#	Model	Year
17	Princeton Reverb	63
18	Deluxe Reverb	63
19	Super Reverb	63
20	Dual Showman	63
21	Vibrochamp	64
22	Bronco	64
23	Dual Showman Reverb	68
24	Bantam Bass	69
25	Music Master Bass	70
26	Quad Reverb	72
27	Super Six Reverb	72
28	Bassman 10	73
29	Super Twin	76
30	Studio Bass	77
31	Fender 75	79
32	Champ II	82
33	Super Champ	82
34	Princeton II	82

Fender Amps

Model 26

The first official Fender amp was the Model 26 which debuted in 1947.

Fig. 59 Front and back of a Model 26,

Champ Amp — Model #1
(also Vibro Champ)

The Champ was the first Fender amp to

carry the Fender nameplate. It was based on the K&F and model 26 amps that preceded it. The first models of the Champ were called the Champion 600, later models were called the Champ 600, and still later models were called the Champ Amp.

The Champion 600 amp was first made in 1947. The covering was two-tone cream and brown lacquered linen cloth which looks like naugahyde. The speaker opening is fairly small with rounded corners. The first tweed Champs came out in late 1948. They were made with the following names printed on the back: Champion 600, Champ 600, and Champ. In 1952 Champ amps received a wider grill opening; the control panel was moved to the top and the name became standardized as the Champ.

In late '53 or early '54, Champs received a dual stage pre-amp tube, and it had an output of 4 watts. The first Champs had a power consumption of about 10 watts, and an output of about two watts. Champs received sculptured grill cloth in 1955, and in 1956 they began to receive 8" speakers in place of previous 6" ones. Champ's tweed was replaced by black Tolex, and a front blackface control plate replaced the older top mounted controls in 1964. There was a change-over period when a few Champ cabinets with top mounted controls were covered with black Tolex. Also, the leather handle of the tweed amps was replaced by a black plastic handle.

Champ amps, because they were economy models, never had all cream vinyl or chocolate vinyl like Bassman, etc.

In 1965 Champs were increased to a full 5 watts, but this was extremely low in comparison to the 100 watt Marshall, Vox, and Fenders of that time. But in the late seventies, musicians stopped relying on mammoth stage amps and began to mike smaller amps with a high power P.A. system. For this reason, low output is not critical because tone is more important, and a little Champ (or other amps like it) if overdriven, can give more than adequate output for most guitarists.

The Vibro Champ was introduced in 1964. Because it has an extra 12AX7 tube, half of which is used for the "Vibro", and half of which adds gain, the Vibro Champ has slightly more power. The Vibro Champ was the first Fender amp not to carry a model number. It would have been model 17 if numbered. In 1982 the Champ was redesigned and became the Champ II and Super Champ.

The Champ was originally designed as a mate for Fender's lap steel guitar. The fact that many of these amps can be found (whereas fewer lap steel guitars are around) is a testament to its value.

In 1982 Fender introduced the Champ II and the Super Champ amps. These amps are very different from simple old Champs. The Super Champ (the more advanced model) has two 6V6 outputs, reverb and three gain controls plus bright and boost circuits.

Fig. 60 Brown and tan Champion 600.

Fig. 61 Wide front tweed Champ 600, with brown jersey cloth grill.

Fig. 62 Black Tolex covered Champ with top/rear controls.

Fig. 63 60's Blackface Champ and Vibro Champ.

Princeton — Model #2 (also Princeton Reverb)

First made in 1947, the Princeton was a three tube amp similar to the Champ and was first advertised as a mate for lap steel guitars. The first Princetons, i.e., pre-date code models, had the name in script silk-screened on the control panel. Princeton is the name of a street five blocks from the Fender company; it was this familiarity that led to the naming of the amp.

The pre-amp tube of the early Princeton was a dual stage 65L7, and not like the Champ's single stage 6SJ7. This made the first Princeton nearly twice as powerful as the Champ, i.e., an output of $4\frac{1}{2}$ watts. When the Champ received a 12AX7 pre-amp tube, the Princeton was similarly upgraded.

In 1957 the Princeton received a "vibro" tremolo circuit, and in 1964 the amp gained separate bass and treble controls. Also in 1963 the Princeton Reverb debuted.

The Princeton went to a solid state rectifier in the mid 60s. After fifteen years of production, the Princeton's wattage was increased from 5 to 12 watts. More importantly, especially in regard to tone, the Princeton began to use a solid state rectifier, and went from three single tubes to six tubes. This cascading of the input contributes to the creation of tonal color. Since there are four dual tubes in a post-mid 60s' Princeton, it is the equivalent of 8 single stages.

The Princeton is thought by many guitarists to be the epitome of "good sound". The Princeton is a classic of amp design and for anyone who prefers the softer sound of a 6V6 (as opposed to a 6L6) the Princeton is nearly perfect.

Fig. 64 Princeton with tweed and TV front.

Fig. 65 Princeton with wide front and diagonal tweed.

Fig. 66 Princeton in brown Tolex.

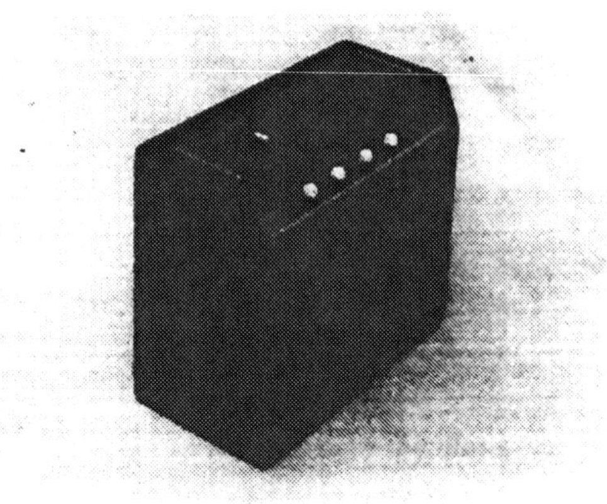

Fig. 67 Princeton in blackface, example is rare early blackface with white knobs.

In 1982 Fender introduced the new blackface Princeton Reverb II. This new Princeton has a 29 watt output, a prescence control, a mid-boost, switchable lead sound, and a 12" speaker.

Deluxe — Model# 3 (also Deluxe Reverb)

The 1948 Fender amp line consisted of of three models: the Champ, Princeton, and Deluxe. With five tubes and an output of ten watts, the Deluxe was the top of the line. Advertisements said "... clean audio power...", guitarists soon found that no other Fender amp could produce as much overdriven distortion. At full volume it is indeed difficult to believe that it is a low power amp.

The original circuit has two 6SC7, two 6V6, and one 5Y3 tubes. The classic Deluxe circuit was devised in 1953, and was composed of one 12AY7, one 12AX7, two 6V6, and one 5Y3GT tube(s). The Deluxe was the first Fender amp with dual output tubes, i.e., the first Fender amp with a push-pull circuit. When Tremolo was added 1957, a 12AX7 tube was added. Half of this tube was the tremolo source and half served as input gain. Originally the mate for the Broadcaster guitar, the Deluxe was superseded by the Super which, in 1950, became the advertised mate of the Telecaster (the new name for the Broadcaster).

In 1964, separate treble and bass controls were added to each of the two channels. This was also the year the Deluxe Reverb amp debuted. In 1982, the new blackface Deluxe was introduced; it is an all new version of a Deluxe Amp.

The tweed Deluxe is one of the most popular older amps. Because many were made, the cost is comparatively low; in 1980 a tweed Deluxe was selling for about $250.

Fig. 68 Deluxe, a late forties TV front model with brown cloth grill.

Fig. 69 Deluxe in brown.

Super — Model #4 (also Super Reverb)

This amp was first made in 1949 and sold as the Dual Professional. It had a V front cabinet with the central strip that was also used for the first Super amp of 1950.

Fig. 70 Early Super amp.

The Super amp debuted with 6SC7 preamp tubes and two 6V6 power tubes, and had about 16 watts. This was the first Fender with two speakers; two 10" Jensens. The dual speaker Super was created when the Deluxe was upgraded to a 12" speaker. Because Fender had a lot of 10" speakers, it was decided to use two 10" speakers to handle the output of the new Super model. The 6SC7 tube was upgraded to 12AYZ, so the 1954 Super put out 20 watts. In 1955 the Super received separate treble and bass controls and it was increased to approximately 30 watts. When tremolo and other changes were instituted, the power was increased to 45 watts. In 1964 the Super Reverb with four speakers debuted.

Solid state rectifiers were installed in 1979, and the power was increased to 65 watts. The changes of the Super through the years reveal that older amps are not necessarily more desirable than newer models.

The Super was marketed as the mate for the Telecaster guitar; a position it held until 1956. After this, it was promoted as the mate to the Esquire. Note: Though the Super, Pro, Bandmaster, Concert, and Vibrosonic amps used virtually the same chassis all through the years, in 1955 the Super was downgraded to 6V6 from 6L6 tubes. Later in 1957 or 1958 the Super was returned to the fold.

Fig. 71 Super in brown.

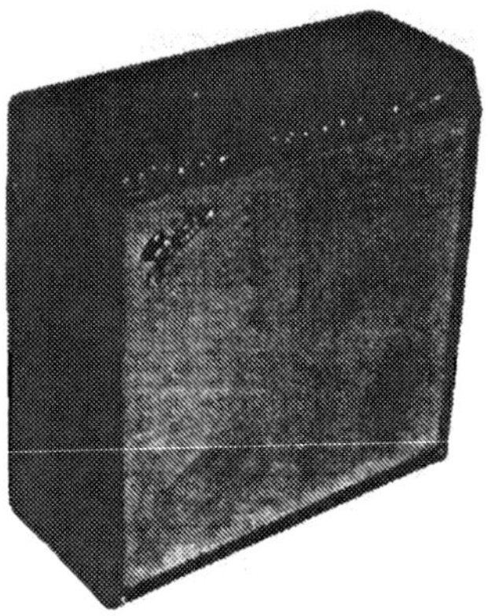

Fig. 72 Super Reverb in black.

Pro — Model #5 (also Pro Reverb)

The Pro is actually a one 15" speaker version of the Super, and it also debuted in 1950. It had three 6SC7 tubes which soon were replaced, in 1954, by two 12AY7 tubes and one 12AX7 tube. The Pro is still made and has undergone the usual addition of tubes for vibrato and reverb. The Pro Reverb debuted in 1964 and had two 12" speakers. Interestingly, the Super Reverb amp is basically a four speaker version of the Pro Reverb. The standard non-reverb Pro still has one speaker.

Fig. 73 Pro amp in tweed with wide front.

Fig. 74 Pro Reverb in black.

Bassman — Model #6

The Bassman amp has probably changed more than any other Fender amp; also, there are many derivatives of it: the Bassman, Bassman 10, 50, 70, 100, and 135. The first Bassman was produced in 1951, the year that Fender introduced their revolutionary solid body electric bass. It is interesting to note that even through the 50s, Fender advertised the Bassman amp as . . . "designed to accomodate string bass".

The very first Bassman had two 6SJ7 pre-amp tubes and a 6N7 acting as an inverter prior to the dual 6L6 power stage. This first amp had only one speaker and it was almost immediately replaced by a Bassman with a 6SL7 inverter and one speaker. By 1955 the Bassman had gained: two 12AY7 and 5U4GA tubes, plus one 12AX7 tube to augment the two 6L6s. At this time the Bassman had an output of 50 watts. A close examination of the 1953 Bassman and Twin amps will reveal that both use the same chassis, but the Twin had dual speakers. In 1954 both received a new chassis with dual rectifiers (used for less hum and more power) but in 1956 when the Twin went to four output tubes, the Bassman stayed with two. In the next twenty years, the only other significant change to the Bassman was the adoption of a solid state rectifier (in 1964).

Fig. 75 Bassman piggyback in white Tolex.

The Bandmaster — Model #7
(also Bandmaster Reverb)

This amp came out in late 1952. It originally had one 15" speaker. In 1954 it was available with three 10" Jensen speakers.

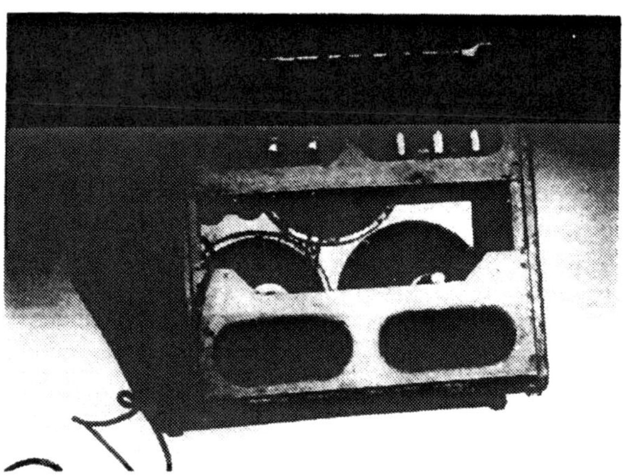

Fig. 76A&B Front and back of early Bandmaster.

The Bandmaster went piggyback in 1961; it had one 12" speaker, and in 1963 it had two 12" speakers and was increased to 45 watts. The Bandmaster amp head is also available with reverb. The Bandmaster line was discontinued around 1972.

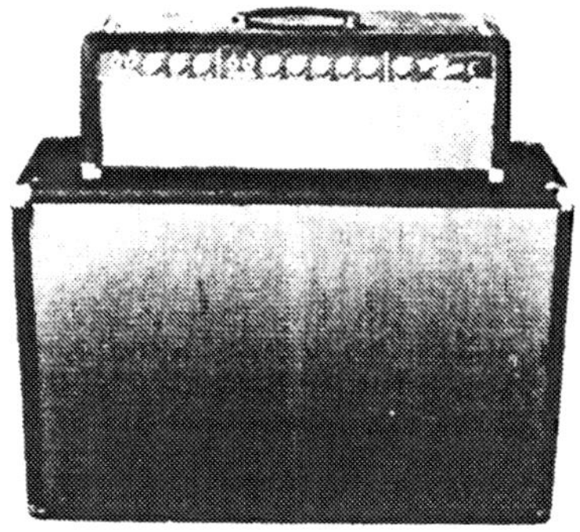

Fig. 77 Bandmaster in black Tolex piggyback.

Twin Amp — Model #8 (also Twin Reverb)

The Twin was created in 1952 as the new top amp of the Fender line. It was initially presented as a mate for the Stratocaster. An interesting note is that the Strat was not introduced in 1953, but rather, in 1954. This is because the original prototype Strat bridge, with adjustment screws exiting the front, was too awkward and the Strat release was delayed until the bridge was redesigned.

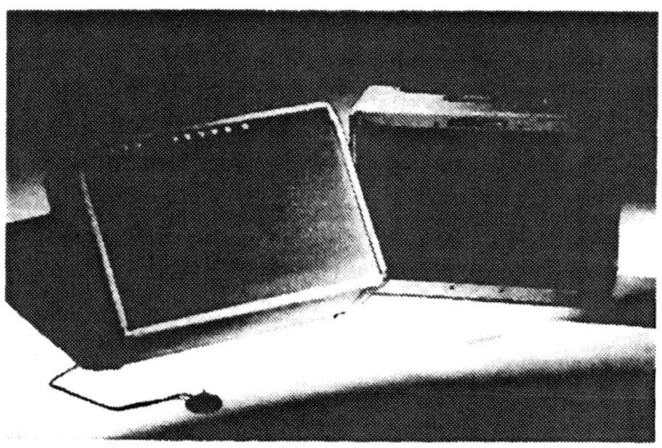

Fig. 78 Twin amps one in white Tolex, one in Tweed.

Fender had decided to make a new amp with two 12" speakers. The first Twin amp, with 15 watts, was born.

The Twin had separate bass and treble controls in 1953, in 1954 it received a presence control, and in 1956 a mid-range. These early Twins had two 6L6/5881 output tubes. In 1954 two 5Y3GT rectifier tubes replaced the earlier 5U4, and in 1955 a change was made to two 5U4GA's. The circuit changed in 1958 to using one GZ34 (a dual rectifier) and four 5881 tubes. The output of the 50 watt 1955 model went up to 85 watts. In 1960 the switch to a solid state rectifier allowed power to be increased to 90 watts. The Twin Reverb was introduced in 1963 and a change in the power transformer (for a higher voltage output) allowed the power to be raised to 100 watts. In 1970 the Twin received a master volume, and in 1973 a push-pull distortion switch. In 1979 the output was raised to 135 watts.

Twin amps have changed greatly over the years. Its evolution is almost phenomenal. A 1952 Twin is more comparable to a 1970 Princeton than a 1970 Twin.

Fig. 79 Twin Reverb in black Tolex & silver face.

Tremolux — Model #9

The Tremolux debuted in 1955 as a 15 watt amp with one 12" speaker; it used a 12AX7 tube for tremolo generation. This was the first use of tremolo in a Fender amp, and it paved the way for tremolo circuits for other Fender amps. On some Fender amps, tremolo is called vibrato. In 1960-61 the Tremolo went piggyback, andd within the following year its speaker cabinet received two 10" speakers. The Tremolux was discontinued in late 1965 because the Vibrolux replaced it.

Fig. 80 Tremolux stack in white Tolex.

Harvard — Model #10

Originally, the Harvard amp was a 5 tube amp; it debuted in 1956. Although Harvard schematics are dated 1955, the amp was not included in the main 1955 catalog. It was advertised as a mate for the Musicmaster guitar. When the Harvard came out, it had two 6V6 tubes, but soon it was downsized to one 6V6 tube. Temporarily, the Harvard was more powerful than the single 6V6 Princeton; but when the Princeton was upgraded, it exceeded the Harvard's performance. The Harvard remained virtually unchanged with the exception of changing 8" speakers to 10" speakers. The Harvard was discontinued in 1960 because there was no need for a model between the Princeton and the Champ. In 1959 Fender did introduce a small economy amp called the Harvard, but this unit is solid state.

Fig. 81 Harvard in tweed.

Vibrolux — Model #11
(Vibrolux Reverb 1964)

Tremolux, with 15 watts and one 12" speakers, came out in 1955 as a mate for the Telecaster guitar. It was the first Fender amp with tremolo. The circuit of this amp was very much like a 1955 Deluxe with the addition of a 12AX7 tube used to provide tremolo. In 1956 the Vibrolux, using the Tremolux circuit, was introduced as a mate for the Duo-Sonic and had a 10 watt output.

The Vibrolux used half of the first preamp tube for tremolo instead of gain boost, so it had less power than the Tremolux. In 1963, both models received the same chassis which now uses two 6L6 instead of 6V6 output tubes; and the Tremolux, originally with one speaker, was available with either one or two speakers. By 1964 the Vibrolux had a partner called the Vibrolux Reverb which had 35 watts, and by 1970 had reached its present output of 40 watts.

In 1967 the Tremolux was discontinued, and the two speaker Vibrolux continued on. The Vibrolux, with its two speakers and dual 6L6 outputs, is often referred to as an economy version of the Twin. In 1964 the Vibrolux Reverb debuted.

The Vibrolux could be described as the "poor-man's Twin" and has always been a good seller for Fender. However, this amp has only 10" speakers; the Twin's 12" speakers give fuller low notes.

Fig. 82 Vibrolux in tweed.

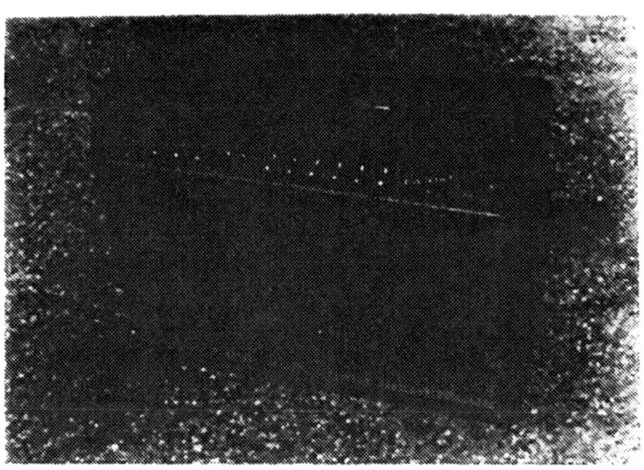

Fig. 83 Vibrolux Reverb in black.

Concert — Model #12

The Concert came out in 1962 and shared the Super/Pro/Bandmaster chassis except that the Concert had four 10" speakers. It's interesting to note that Fender offered the same amp in four versions: the one speaker Pro, the two speaker Super, the three speaker Bandmaster, and the four speaker Concert. In 1967 the Concert was discontinued. (Note: the Super Reverb does have four speakers.) in 1982 the Concert was revived.

Fig. 84 Concert in black.

Vibrasonic — Model #13

The first Vibrasonic, model 5G13, debuted in 1957. It had about 25 watts, a 15" speaker, and was similar to Pro, but with the addition of vibrato. It was superseded by the Vibroverb, which debuted in 1962. The Vibrasonic was discontinued in 1967, but reborn again in 1982 as the Vibrosonic Reverb, an amp sharing the Twin Reverb's chassis.

Fig. 85 Vibrosonic Reverb, a 1982 revival.

Showman — Model #14

Single Showman, Dual Showman, Dual Showman Reverb

The Showman line was Fender's real "showpiece". These were high-power amps made to equal the Marshall and Vox amps of that time. The obvious differences between the Single and the Dual was that the Single had one speaker and the Dual had two. The two used in the Dual was a JBL130E (a 12" speaker), and a JBL140 (a 15" speaker).

The Single was the first Showman and it came out in 1961 with 85 watts. It was available as the Single Showman 12 with a 12" speaker or the Single Showman 15 with a 15" speaker. The Single Showman 12 was discontinued in 1966, and the Single Showman 15 was discontinued in 1968.

The Dual Showman debuted in 1963, and used the same basic circuit as the Single; i.e., four 6L6 power tubes. The Dual Showman Reverb came out in 1969, and about six months later, the plain Dual Showman was dropped from the line. Note: The Showman line has always used the chassis of the Twin. So it is practical to think of the Showman as a piggyback version of the Twin.

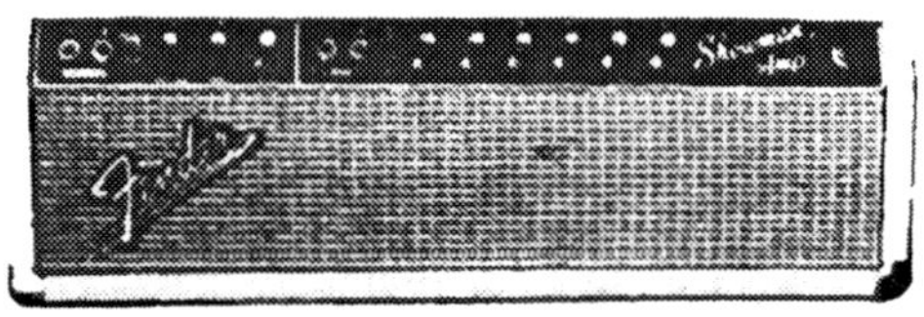

Fig. 86 Showman head in white Tolex.

Fig. 87 Dual Showman Reverb silverface stack. Note, speaker cabinet is on its end and nameplate has been repositioned.

Reverb — Model #15

Though Fender was not the first company to make a reverb unit, Fender's reverb was the first unit to achieve widespread popularity.

Fender first offered reverb in a self-contained unit in 1961. An instrument had to be plugged into the reverb unit and the output of that unit had to be plugged into an amp. In 1962 Fender began to offer reverb built into many of its amps. For instance, the Super amp was offered in two models, the Super Reverb amp (with reverb) and the Super amp (without reverb). Because tweed was discontinued when reverb was instituted, no Fender tweed amp had reverb.

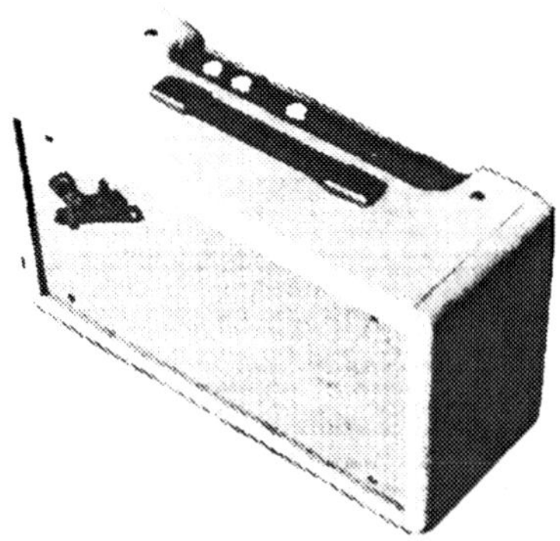

Fig. 88 Reverb unit in white Tolex.

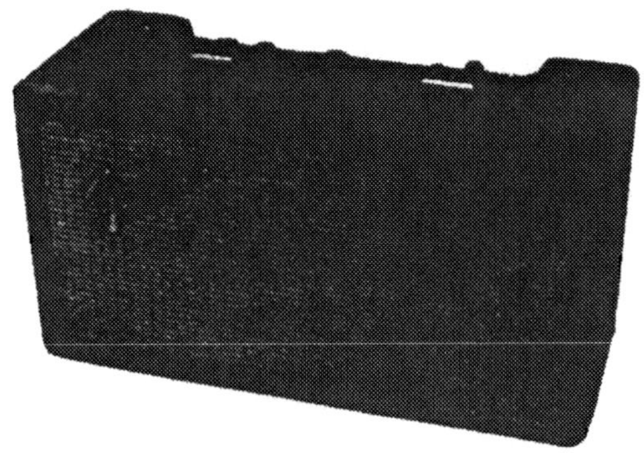

Fig. 89 Reverb unit in brown, later ones were in black.

Vibroverb — Model #16

This amp was introduced in 1962 as a reverb version of the Vibrolux. In 1967 it received two speakers. The Vibrolux Reverb debuted in 1964 with two speakers. In 1967, the last year of the Vibroverb, it was available with either one or two speakers. It was discontinued because it duplicated the Vibrolux Reverb.

In 1964 some Vibroverbs went to one 15" speaker. Throughout Vibroverb's history it appears buyers could choose speakers.

Fig. 90 Vibroverb in white Tolex.

Bronco

The Bronco amp was actually identical, at least inside, to a Vibro-Champ. It was first made in September of '64 and discontinued in the early 70s.

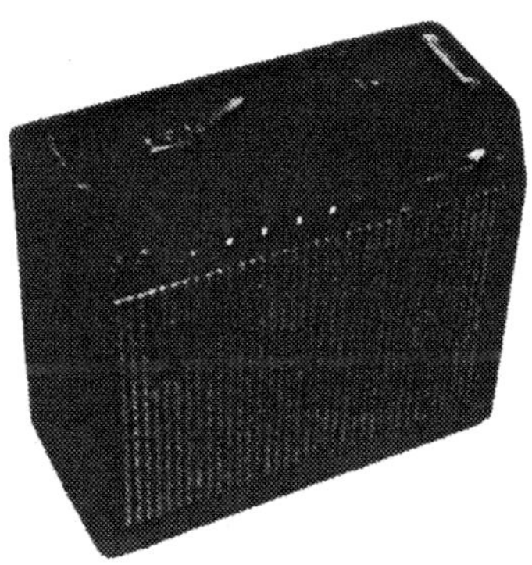

Fig. 91 Bronco in silverface, name Bronco is in red instead of usual Fender turquoise.

Bantam Bass — 1969 – 1972

The Bantam Bass is a smaller version (one pre-amp tube less) of the Bassman. It was designed around a trapazoidal plastic cone speaker bought from Yamaha. The

Fig. 92 & 93 Front and back of Bantam Bass.

poly-plane speaker was too hard for Fender to obtain, so the Bantam was discontinued. The Bass 10 uses a conventional speaker and was derived from the Bantam.

Super Bassman

The Super Bassman is similar to a 1960s' Twin amp chassis with two inputs: one for bass and one for guitar. However, the Super Bassman did a different output section and transformers.

Music Master Bass — 1970

This small tube bass amp used two 6AQ5a (10 max watt) output tubes. Because these tubes became unavailable, Fender switched to 6V6 tubes. The older version had a 12 watt output, whereas the newer models had about 18 watts.

Quad Reverb/Super Six Reverb — 1972

These two amps used the Twin chassis in a large cabinet. The Quad has four speakers and the Super Six has six speakers.

Super Twin -- 1976
(also Super Twin Reverb)

This new design differs from all other models. It uses six 6L6GC output tubes to provide 185 watts, making it Fender's most powerful tube guitar amp available.

Studio Bass — 1977 - 1981

The 200 watt Studio Bass is the bass version of the Super Twin. The chassis is the same except for the tone control section. It is one of the most powerful tube bass amps made.

Note: Fender has made many other bass amps, but nearly all have been of solid state design.

Fig. 94 Fender group shot of late 1970's models.

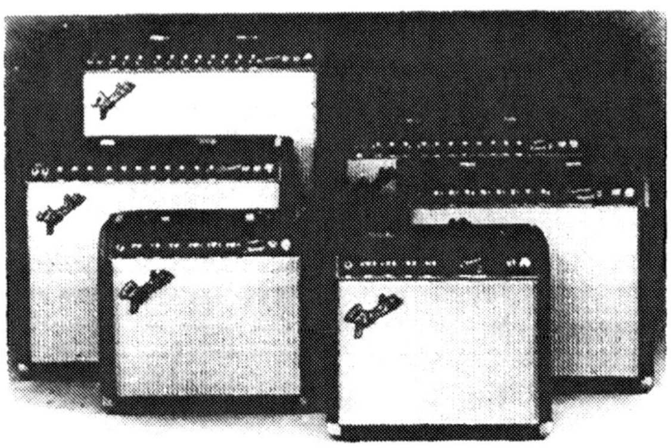

Fig. 95 Fender group shot of early 1980's models.

White Amps

In 1955 Leo Fender created the White brand amps as a surprise tribute to Forest White who had done circuit work for Fender Company. These amps were sold in conjunction with a White lap steel guitar, and internally comparable to Fender Princeton amps made at that time. The label inside has a schematic, which reads 40 watts, but this denotes power consumption, not output. RMS output was about 5 watts.

The outside of White amps had blue-gray linen tweed which resembled plastic cloth. This grill cloth was very dark blue with blue dyed leather handles. The label read "White Instruments, Fullerton, California", and approximately 1,500 of these models were made.

Fig. 96 White brand amp.

Fender Amp Model Dating

The method of dating a Fender amp is as follows: Starting in 1952, there is a code designation inside the amps. The early code consists of a number, followed by a letter, followed by a one or two digit number. For instance, 5C1 would translate as follows: the 5 represents the decade of the 1950s; the C stands for the fourth year of 1950 (1953), and A stands for 1, B for 2, C for 3 and so fourth. The remaining numbers in the code stand for the amp model number.

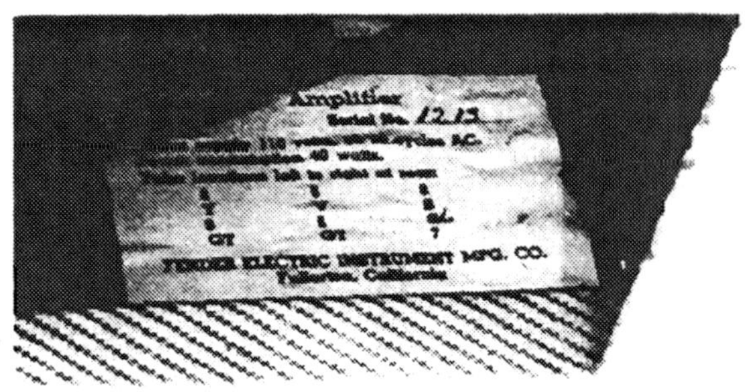

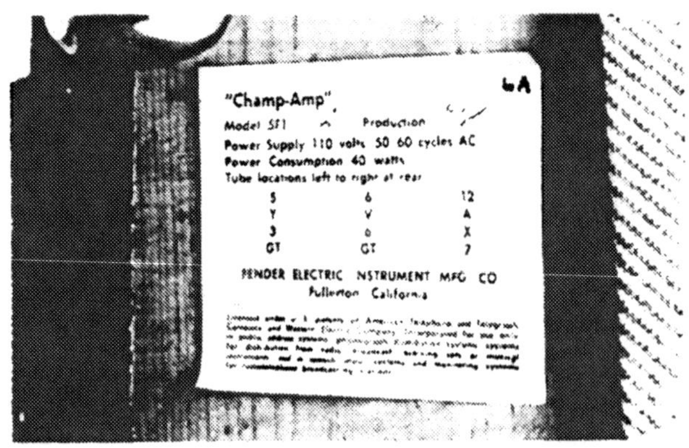

Fig. 97 & 98 Pre-dated code label and a (1956) date code label.

Old undated schematics which do not have the number/alphabet code do exist. In some cases, one or two circuits predated coded circuit schematics. This coding procedure was not created until 1951 and, for this reason, there are no Fender amps with a 5-(1950) designation, and very few with 5A (1951). Therefore, Fender amps made from 1946 to 1950 have no date code. The code was not fully implemented until 1953 when labels inside amps carried the code of 5C followed by the model number. Schematics labled with the 5C code are the earliest coded models of the Fender Company.

In 1960, the code for a Champ could read 6F1, meaning 1966 model #1. The Vibroverb, model #16, was first made in 1963, and the first production was labeled 6C16. In 1963 a decision was made to use a new coding method. This code consisted of two letters followed by three numbers. A typical example is AB763. The first letter A stands for amp. The second letter stands for circuit version. In the case of B, this would mean a second version. The first number, in this case a 7, would mean the seventh revision of that version. Two digits would represent the year, e.g., 1963. Note that the name of the amp had to be used with this code. After 1963's model #16, numbers were no longer applied to differentiate models.

GIBSON

In 1935 Gibson came out with its first electric instrument (the EH-150) which was an electric Hawaiian guitar. Shortly thereafter, they came out with an electric Spanish guitar (the ES-150), and electric banjos (the EB-150). At this time, Gibson also announced a line of amps primarily designed for the electric Hawaiian guitar. These first amps were named the EH-275, EH-150, and the EH-185;[1] they were made by Lyon Heally for Gibson. The wattage ranged from 8 to 20 watts, and all models had D.C. field coil speakers made by the American companies — Rolo or Jensen. The EH-185, played by Charlie Christian, was the most popular of the three. This system had a hinged lid which allowed the amp chassis to lift out; when closed, it functioned as a bass reflex speaker. The speaker itself was 12", and the amp section had two metal case 6L6 output tubes producing 17 to 20 watts. The EH-150 was an open back model with a 12' speaker.

Fig. 99 An EH-150 amp made between 1935 and 1938, later models had rounded, non-protected top corners.

[1] The EH-185 was first made in 1935 according to some ex-Gibson employees, but others say the date was 1957. In any event, Charlie did play an early Gibson amp.

There was also an EH-100, but this amp is not depicted in catalogs. It looks so old that it is very possible it is the very first Gibson amp. No one from Gibson knows for sure.

Fig. 100 An EH-100, a very old example.

Seth Lover, inventor of the Gibson PAF humbucker, was largely responsible for the design and development of the Gibson amps. After the war, the Gibson staff created the BR amp series. The BR stood for Barnes Reincke, the designers of the amps. The amps were made for Gibson by Magnavox in Indiana. The BR-6 was a small amp, and the BR-9 was a larger model. But because the BR amps still had field coil speakers, they suffered from annoying hum produced by the speaker's power transformer.

Gibson's GA amps, which came out in 1950, were the most popular tube amps. They were the first Gibson line amps to have permanent magnet speaker (although there had been some experimentation with other models). The GA-40s were first issued in late 1952 and was known as Les Paul amps. The first models had plastic script letters (LP) on the grill cloth. Later models had a small oval plate at the center bottom of the grill cloth which read "Les Paul". This amp sold as a companion to the Les Paul guitar. The GA-5, similar to the Fender Champ, was sold as a companion to the Les Paul Jr. guitar.

Fig. 101 First a BR-6 Ultratone, top model of the original BR line. On the right is the BR-9, later the top BR model.

Fig. 102 Les Paul amp of 1952, later models don't have the LP design.

Fig. 103 Les Paul Jr. amp, GA-5. In the late fifties, GA-5 amps had square cabinets with wide grills and carried the Skylark name.

There are two confusing points about the GA series: 1. Generally speaking, the lower the model number, the lower the output wattage; (i.e. GA-5 has less output than a GA-40; however, a GA-6 has 14 watts, whereas a GA-8 has only 9 watts. 2. The numbers are not necessarily chronologically ordered; i.e., the GA-20 came out in 1954, but the GA-19 came out in 1961.

The following is a brief guide to the G.A. series:

GA-5 "Skylark" $4\frac{1}{2}$ watts (3) tubes
(1) 8" Jensen
GA-5T (w/tremolo)
GA-6 "Lancer" 14 watts (5) tubes,
(1) 12" Jensen 4/9/57
GA-8 "Gibsonette" 9 watts, 4 tubes
(1) 10" Jensen
GA-8 T (Tremolo added)
GA 9
GA 14
GA 15
GA 17 RVT "Scout" (6) watts.
(1) 10" Lansing 1963
GA 18T "Explorer" 14 watts, (5) tubes,
(1) 10" Jensen 1959/GA18T '62
GA-19 RVT "Falcon" (7) tubes,
(1) 12" '61
GA-20T "Ranger" 16 watts, (7) tubes,
(1) 12" Jensen 8/5/54
GA-25 RVT " Hawk" (7) tubes,
(1) 15"
GA-30 "Invader", 16 watts, (6) tubes,
(1) 12" and (1) 8" Jensen + 30RVT
GA-40(T) "Les Paul" 16 watts, (7) tubes
(1) 12" Jensen 8/5/54

Fig. 104 GA-38, made in the mid-fifties, this amp led to the more popular twin speaker GA-25.

The 1960's were the space program years, and the powerful multi-stage rocket projects used in the U.S. space effort were the Atlas, Mercury, and Titan. Consequently, in 1963 Gibson came out with a line of multi-staged, rocket-named amps — the Power Plus line. These amps were the first Gibson amps that had separate amplifier heads and speaker cabinets. The six models, were all designed by Dick Evans, Gibson's chief engineer at that time. Similar to the Power Plus amps were the Medalist line. These amps were self-contained (i.e., not piggy-back units). This line included the Titan Medalist, Mercury Medalist, and Atlas Medalist.

Fig. 105 GA-25 Hawk amp with two speakers.

Fig. 106 GA-8, an early model.

Fig. 107 GA-8T, a late model.

Fig. 108 GA-400, sometimes called Super 400.

Fig. 109 Titan amp stack.

Gibson also made a line of tube amps issued under the Epiphone name. Internally, this EA series of amps was the same as Gibson's, but with different cabinets and coverings. The highest powered EA amp had 50 watts. The EA line was introduced in 1961 and discontinued in 1965.

Gibson acquired Epiphone in 1957. Before this date, the Epiphone Company made guitars and amps that carried the Epiphone name. Actual Gibson/Epiphone guitar production began in 1959, yet references to Gibson/ Epiphone amps were not seen until 1961.

Around 1968, Gibson began offering 50 watt (TR-500) and 100 watt (TR-100) solid state amps. The TR-100 as the TR- 100 RVT, was available with reverb, vibrato, and tremolo. Also in this time period, Gibson introduced a smaller solid state line of amps known as "Maestro". The main models were the M1-RVT, M2-RVT, M201, M202, M204T, and M208T. This line was moderately successful for Gibson, but was phased out around 1967. It was at this time that Gibson began the famous/infamous SG series of solid state guitar amps.

Fig. 110 Epiphone amp, made before Gibson bought the company.

The relative unsuccess of Gibson tube amps could be explained by a quote from a 1963 Gibson catalog: . . . " Gibson amplifiers provide more and finer sound without distortion. . ." Because Gibson amps were more sophistocated than its competitors', the sound they gave was clean and mellow; however, many young guitarists preferred a brash raw power with a hard, driving sound. The sound of Gibson amps is a bit delicate and refined.

MAGNATONE

The guitar introduction to Buddy Holly's "Words of Love" certainly has a magical quality. The reason that the sound isn't heard any more is mainly due to the fact that few guitarist's have an amp like Holly's Many guitarists have the old Strats like

Fig. 111 Magnatone amp, a small, older model.

Fig. 112 Magnatone amp, a model 280.

Buddy's, but who has a Magnatone amp with its shimmering liquid glass vibrato? Note: Buddy Holly also used Fender amps as well as multiple over-dubbing, so on some cuts it's not the sound of just one Magnatone.

The Magna Company was established in Los Angeles in the late 1940's by Paul Barth, the co-inventor of the first practical electric guitar. The Magnatone name originally was a tradename used on amplifiers and instruments sold by Montgomery Wards. By the mid fifties, Magna was producing and marketing an extensive line of Magnatone amps. The amps ranged from AC/DC powered "Starlet" with about 3 watts, to the M-13 "Imperial' with 80 watts. Magna made over 16 models of amps.

The early 1950's Magnatone line consisted of the following models. All models were covered with tan/brown Tolex-like material. This series of amps launched Magnatone's fame. One reason for this was that around 1955 Magnatone changed from dual stage 12 volt tubes to single stage 6 volt tubes.

Earlier Magnatone name amps were contracted from Valco. Barth's Company was located in Santa Ana, and the original name was Bartel Guitars.

In the early 1960's, Magnatone completely redesigned its line. The amplifier's cabinet tops were formed of molded Royalite purple plastic. The amp chassis was mounted to the underside of the top, and could be lifted out of the amp and turned upside down, so that all the heat would rise away from the amp. Although these amps did have a few transistors, their use was limited to the reverb section. A very interesting feature of these amps is that they had true pitch changing vibrato, not amplitude changing tremolo that is called vibrato. This is the reason that other "vibrato" amps can never give the sound Holly had. At a time when other companies were using cheaper magnets, Magnatone continued to use non-ceramic Alnico 5 magnet speakers. These speakers were supplied by large, heavy-duty output transformers. Looking back twenty years and more, it seems seems a bit strange that Magnatone catered more to accordion players than to guitarists. Near the end of its existence, Magnatone went from being a low-medium priced, quality amp, to being a premium amp recognized by pop musicians. However, it was too late.

AMPEG

Although Ampeg is most famous for its monster amps, they did have low wattage amps as well. When guitar amps were first made, the most powerful amps were no more than 15 or 16 watts. The following is a guide to Ampeg amps.

Ampeg 12 (J-12, M-12, TR-12, R-12R)

This student/practice amp was available in several styles. The basic model was the J-12 "Jet". A souped up version was the R-12 Rocket which was like the Jet but had tremolo, extra input, and the cabinet was 2" taller. The R-12R "Reverbrocket" was like the R-12,but had reverb added. Another version of the Jet (J-12) was the two-channel version, the M-12 Mercury. Originally a tube amp, in 1966 it went solid state.

ET-2 Super Echo Twin

This was a 30 watt stereo amp, and each channel had its speaker in the single amp cabinet. Originally tube, in 1965 it went solid state.

Ampeg Portaflex Amps

The Ampeg Portaflex bass amps are designed after the oldest National and Gibson designs. Attached to the top cabinet board is a shock mounted amp. This board is flipped over and latched down so that the amp is sitting on top of the cabinet. This arrangement does several things: 1. Allows heat from the tubes to rise up and away from the amp. 2. Removes the tubes from the direct path of the speaker's sound wave vibrations, and therefore, prolongs tube life. 3. Creates an efficient bass/relfex speaker cabinet. 4. Keeps the speaker from being heated by the tubes. The Portaflex range consisted of the B-12N, B-12X, B-15N, B-15X, B-18N, and B-18X. Because these amps weighed 50 lbs., for the 15 watt B-12N, they were at a disadvantage compared to solid state units that weighed about half that amount. Also, most bass players today want at least 100 watts for even small gigs. Ampeg introduced the solid state BT-15C Portaflex 100 watt bass column. This unit weighed 82 lbs. and had much more practical appeal to bass players.

SVT

When it comes to powerful amps, the name Ampeg is held in awe. The Ampeg SVT with its 300 RMS watts/700 watts peak, and its 295 pound total weight, makes many guitarists faint of heart. The SVT offers unheard of power in a tube amp, and was designed for musicians who need concert hall levels of sound. The SVT system is composed of three parts: A 14 tube amp head weighing 85 pounds, and has two 110 lb. speaker cabinets — each with four 10" speakers.

The SVT has very complex circuitry; a necessity to produce an output higher than any other tube amp made. Since the power levels of this amp are so high, great care needs to be taken when servicing it. The SVT has a current flow of 15 amperes, and this can cause fatal shock hazards.

SST and SBT

These solid state Ampeg amps resembled the SVT. Each had three cabinets; one amp head, and two speaker cabinets. Each SST speaker cabinets had four 12" Altec-Lansing 417B speakers. The SBT had one 15" Altec Lansing 421A speaker in each cabinet.

Fig. 113 Ampeg SVT stack.

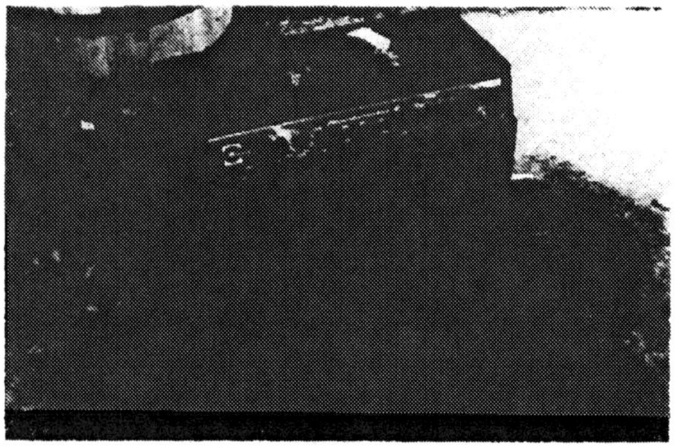

Fig. 114 Ampeg Reverb Rocket.

VOX

Vox has probably had the most complex history and product diversification line than any other musical instrument company. Because of this, writing about Vox and its products is difficult and confusing.

Vox amps were first made by Jennings Musical Industries at 100 Charing Cross Rd. in Dartford Kent, England. Tom Jennings' involvement with amps began when he made one of the first portable organs — a single manual unit with six drawbars. This instrument, with tones similar to a Hammond organ organ, was called the "Continental Organ". In 1957 Tom came out with a tube amplifier called the Vox AC-15. The AC-15 had a power of 15RMS watts and came with one 10" Celestion speaker. This amp, though small by today's standards, was the most powerful amp around at that time. The English group, the Shadows, took these amps on their first tour. Because of these tours, and those of other groups, Vox amps became popular throughout Europe. The AC-15 was discontinued in 1965.

In the late 1950s Vox was a simple, new, small company that made musical instrument amplifiers. They were a "working man's amp" company. Because of import duties and shipping expense, Vox had the U.K. market almost to itself (just like Fender had the U.S. market). In England, and the rest of Europe, Vox was not unduly expensive — Fenders cost much more; therefore, Vox was a very popular seller.

At this time there was a world-wide surge in pop music interest. One English group using AC-15 amps was about to create far-reaching effects on music and musical instrument merchandising. When the new high-power Vox amp, the AC-30 (30 RMS watts and one 12" Celestion G-12M speaker) came off the drawing board, this club band switch switched to these new amps. They also switched their band name from the Silver Beatles to the Beatles, and the results were . . . well, needless to say: "The rest is history". But this was actually only the "birth" of Vox.

Tom Jennings was in charge of Vox's outward design, selection of components, and overall quality control; Dick Denny was responsible for the design of Vox's amplifier circuits. Denny's work started with the AC-30, the Vox line of England, and the Vox line later to be made in America. Together Jennings and Denny designed very good looking amps with a unique, strong treble tone.

When the Beatles and other musicians performed, people saw the Vox gear behind the group; and as a result, there was a tremendous demand for Vox products. In the sudden explosion of music, Vox used music groups for endorsement purposes, and groups used Vox to establish credibility. The legend of Vox was both spontaneous and manufactured. No one was immune to realizations of commercial potential, and Vox grew tremendously.

The year 1964 was action-packed for Vox. It came out with the AC-100, the most powerful amp in the world at that time. This was a 100 RMS watt tube unit consisting of a separate amplifier head and a speaker cabinet with four 12" Celestion G12M speakers. This amp, called the Supreme in England, was used at the Beatles' Shea Stadium concert. It is claimed that these amps were hot-rodded to 200 watts each, but the crowd of 20,000+ created so much noise that the Beatles could not be heard very well.

The year 1964 was momentous for another reason. Tom Jennings became friends with the Thomas Organ Company located in Sepulveda, California, and an arrangement was made whereby Thomas became the importer of English Vox equipment.

Vox was now a very different company than nine years earlier. It produced a line of amps from 15 watts to 150 watts, a line of guitars including over thirty models, and eventually sold drums, lighting equipment, and a complete line of band instruments (due to a tie-in with King Instruments). More important, however, was the roster of endorsees.

Vox stepped into the 1960s music scene much more aggressively than its competitors. The individuals and groups using Vox equipment was staggering; from Lawrence Welk to the Beatles; from Waylon Jennings to Buddy Rich. It seemed that the whole music world was using Vox.

As audiences became larger and began to scream more loudly, more power was needed by musicians in order to be heard. Now companies like Vox made more powerful amps. They put more power tubes in a small cabinet, etc., but there was not sufficient cooling for use during long performances, and the components were not substantially over-engineered. (This was especially a problem for the AC 100). This resulted in many rock concerts being abruptly halted because an amp died in a cloud of smoke. It is interesting to note that at this time, The Who developed their act of throwing their guitars into exploding amps. The group did use Vox equipment, and it is a plausible assumption that their stage act developed as a result of frustration with their equipment.

The problem of tube amps overheating caused many companies (and Vox was no exception) to switch to transistors, which promised greater reliability. Solid state was seen as a glamorous panacea. Transistors had the status of being space age modern. It was Vox of England, not Vox/Thomas in America, who initiated the idea of solid-state Vox amps. This switch to transistors was also prompted by the fact that Jennings sold Vox to the Royston Group in early 1964. Jennings stayed on with Vox as the Chief Executive Officer until 1969. But it was Thomas/Vox of the U.S. that had the expertise and facilities suited to manufacturing an entire solid-state Vox line. Dick Denny went to Thomas in Sepulveda, California, to help create the new Vox line that debuted in 1965. In time, Vox in England also had its own line of solid-state amps.

Solid state was seen as the means to hold on to Vox's enormous stable of star musicians, and despite the different sound of solid state, many musicians (including the Beatles) remained Vox enthusiasts. The Vox popularity was so strong that people were willing to spend great sums to possess something "Vox". The price of a 100 RMS Vox Beatle amp in 1965 was over $1,000. Considering inflation, that is equivalent to about $2,500 in today's market. A new 100 RMS can now be purchased for $700.

The solid-state line did offer some advantages; the tonal circuits were more reliable, and the solid-state equipment weighed much less than tube gear — a fact all roadies appreciated. However, it did have some problems as well. The germanium transistors were not reliable, and the T-60 bass amps suffered from bias and thermal run-away problems by virtue of its transistors and its direct-coupled, split-rail circuit. This amp became the most notorious Vox model. Vox's switch to silicon transistors did improve the reliability of other Vox models.

One interesting feature of Vox solid-state design is the use of an FET component pre-amp in some of the models. An FET helps to preserve the warm sound of a guitar and doesn't introduce harsh bi-polar transistor-type tones into the music. The majority of amp companies now use this component.

In 1968 rumors came out which stunned the music world; the Beatles intended to disband. It seemed that this was a portent of rough times ahead both for music groups and manufacturers of music equipment. In 1968 Vox of England was sold to Erith kent, and in 1969 Jennings left as Vox's CEO. Jennings went off on his own to form Jennings Electronic Industries Ltd. He produced Jennings Amplifiers for about two years. These amps were solid-state and used slide rather than rotary potentiometers.

Meanwhile, things were equally difficult for Vox/Thomas in America. Because of cost and reliability problems, as well as the desire for the original sound, musicians began to leave the Vox camp. Even Thomas' flamboyant Vox Music Scene magazine could not boost Vox's waning sales.

The music business' downward spiral helped contribute to Vox's decline. Vox was resurrected in 1980 and today they produce tube and solid state amps. The tube units are faithful to the 60's Vox tube amps.

Fig. 115 VOX AC-30 combo.

Fig. 116 VOX AC-30 head.

MARSHALL

Because Marshall is a comparatively young company, not very many models have been made.

In 1965 Jim Marshall, a drummer who owned a music shop near London, put together a fairly simple but powerful 100 watt amp for The Who. The idea was to made something that would look and sound powerful. This original Who amp was composed of the amp and eight 12" speakers in one huge cabinet. For the sake of portability, the unit was divided into three parts: the amp head, a cabinet with four speakers, and another cabinet with four speakers. The circuit used for that amp was not that common, i.e., it used a cathode follower, and as typical with these designs, had a fairly low impedance circuitry. It is interesting to compare the similarities of early Marshall 50 watt amps and Fender 4/10 Bassman amps.

The distinctive sound of a Marshall, one of zingy presence, was due to two major factors: 1) the placement of the tone controls within the circuit) relative to the gain stages and the pre-amp section, greatly affected the tone; and 2) the use of the fairly light duty easily overdriven Celestion speakers. To achieve sustain, the amp must be turned up to full volume. This fact has caused a bit of a problem because many guitarists don't always need that kind of power. The use of power attenuators wasn't completely successful, and for this reason, Marshall included the master volume in its later amps. Another factor influencing Marshall sound, is the fairly light-duty output transformer; this is often difficult to notice because many Marshalls have had their original transformers replaced (because of failure).

The Marshalls used by many famous guitarists have not been stock units. Though EL-34's are the standard Marshall 100 tubes, Hendrix sometimes used KT66 or KT88s in his units. His hot rodded high output amps were delicate and temperamental. The units often burned up after a short period of heavy use. Like most other companies, Marshall's production models are a compromise between short-lived blazing fireworks of sound and dependable mundane performance. But a stock Marshall 100 watt head is no slouch, it can put out 170 watts when run in tandem.

Marshall amps sold in the U.S. often have 6550 tubes instead of EL-34 tubes used in U.K. Marshalls. A repairman can change the fixed ballast resistor so that a U.S. model can take EL-34s. Another alternative is to use 6AC7 American tubes which have a hard brash sound instead of U.K. EL-34 which have a softer, smoother sound.

Model #1959

The 1959 name of this amp does not refer to a year, it is simply a model number. This 100 watt head is as close to the 1965 Marshall head as any made by Marshall.

Model #1987

This is a 50 watt lead/keyboard amp head.

Model #2203

This 100 watt head for guitar has switchable impedances, pre-amp volume control and master volume controls as well as other features.

Model #2204

This amp head has the same features as the 2203, but with only 50 watts.

Model #2195 & 2959

These models are solid state units.

Model # 1986

This model has a 50 watt bass head.

Model #1992

This model has a 100 watt bass head.

Marshall also makes several combos. The Rock and Roll Combos are solid state, but the Country Club Combos are tube units. The 4140 and 4145 are 100 watt guitar amps, and the 4150 is a 100 watt bass amp.

Marshall Speaker Cabinets

The traditional Marshall stack (a stack is a unit of an amp and speaker cabinets) consists of an amp on top of one or two Marshall speaker cabinets. Most Marshall cabinets use four 12" Celestions. The #1935 cabinet differs from the #1960 in that the former has more pliable cones which extend the bass frequency response of the lower portion, making this unit well suited for bass players. The stiffer cones of the #1960 are suited for lead guitarists. Marshall does offer JBL speakers as an option, and this sound is much cleaner than that of Celestion's.

Marshall makes two styles of four speaker cabinets. One has a flat front (models #1960A and 1982B); and the other (models #1960A and 1982A) has an angled front which projects the sound of two speakers a few degrees over a crowd.

Fig. 117 Marshall 100 watt head, 1960's style.

Fig. 118 Marshall 100 watt stack, 1980's style.

Smaller Amp Companies

ALAMO

The Alamo Company is primarily known in the Southwest part of the United States, and had its heyday in the 50s. Its instruments and amplifiers quite closely followed those of Fender. The basic Alamo amps looked like Fender amps, but with gray and silver fabric. Alamo did offer special coverings: One was the "Fiesta" look, fabric speckled with bright colorful dots; the other covering was the gray and blue two-tone, used on the little "Jet" model.

Although old Alamo amps have an enthusiastic following, others feel the amps lack a full tonality as well as reliability. This is a condition that effects nearly all fairly obscure and older amp brands.

Fig. 119 Old, pre-Tolex style, Alamo amp.

DANELECTRO

Danelectro amps are as legendary as the Danelctro guitars. However, the amps are more advanced than the guitars.

Most Danelectro amps have a control panel in the upper portion of the back and the aluminum chassis is on the bottom. This helps prevent the tubes from overheating.

Fig. 120 Danelectro DM-10 amp.

The tonal quaility and volume of sound produced from Danelectro amps can vary a great deal. If an extremely bad-sounding Danelectro is encountered, it is probably a case of bad or worn out amp components.

Danelectro made many amp models, the most well known are those made for Sears which carried the Silvertone brand name. These amps do not carry the Danelectro name; therefore, to detect a Sears/Danelectro, look for an aluminum amp chassis which is often at the bottom of the amp cabinet.

Danelectro did make and sell under its own name as well as that of Silvertone, a combination guitar amp/guitar case. These units look rather simplistic and crude, but the sound fairly good, even with the tiny stock speaker.

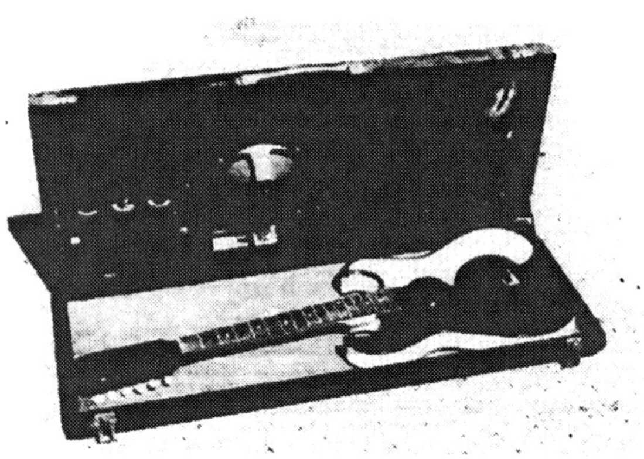

Fig. 121 Danelectro amp/guitar case outfit.

GRETSCH

Early Gretsch amps were made for Gretsch by Valco and other companies. Gretsch tube amps are unique because they were oriented more to country music than to rock.

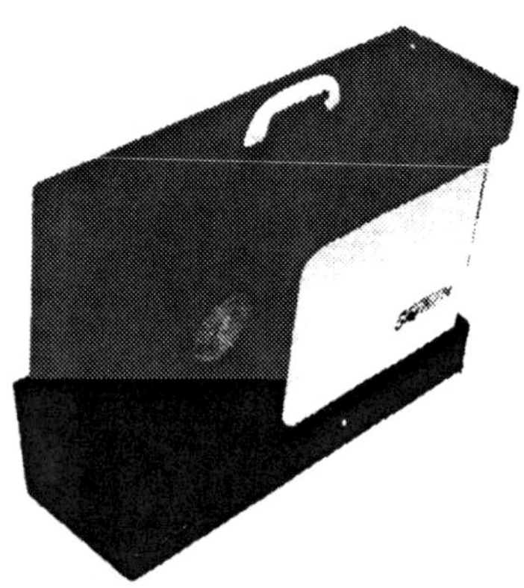

Fig. 122 Gretsch amp.

Chet Atkins Amp

This 1959 to 1964 amp of 37 watts has a special Super Treble Brilliance circuit as well as treble and bass boost circuits. The clean, warm tone came through two 12" Jensen speakers. The configuration of this amp resembles the 1950s' Fender amps; it has a control panel located on the back edge of the top.

Gretsch also offered many other tube amps, including bass and stereo amps.

STANDEL

Standel amps reached the height of their popularity during the 1960s' California surf music craze. Up until 1964, Standel amps were all tube units. In 1964/65 Standel introduced hybrid amps, i.e., amps with six transistors for the pre-amp, and tubes for the driver/power section. In 1966 Standel switched to all solid-state.

Standel, like many other companies, listened to electronics engineers, not musicians. Their press releases state that tubes are delicate, unreliable, and cause distortion; whereas transistors offer durability combined with less harmonics and intermodulation distortion. Standel solid-state amps, like virtually all others, did not meet financial success because musicians did not like the sound as much as other amps.

The early Standel amps consisted of models with names that were retained when the switch was made to solid-state. For example, a 1963 Artist XII is all tubes, a '64 Artist XII is half and half, and a 1966 Artist XII is all solid-state. The early models, however, are not hard to differentiate. Old Standels have medium gray Tolex, a dark grill, a handle on top, and the cabinets are wider than they are tall. 1966 Artist XII are taller than wide, have black Tolex, and handles on the side. The Artist was available with a 12" ceramic magnet Jensen speaker (Artist XII) and with a 15" speaker (Artist XV). The Artist line had combos and piggyback models. The piggyback units names were: Artist XII, Artist XV, and the Artist X (with two 10" Jensens).

Other Standel amps were the Custom line and the Imperial line. The Custom and Imperial amps were internally similar, but the Imperial, which was the top of the Standel line, had more speaker power.

New Tube Amp Revival Companies

ACOUSTIC

The Acoustic Company makes both solid-state and tube model amps. They are very up-front about influences. . . "Boogie started the small size, high power amp [and] everybody seems to have their own Boogie type amp."

Acoustic's "Boogie" (Model G100T-112) is a 60/100 watt tube amp in an oak and walnut cabinet with a dark brown fabric grill. This tube amp has a 4 tube power section connected to a switch that can put two of the tubes on standby via the use of cathode grounding. This means that the amp can be used as a 60 watt model when the use of the full 100 watts would be too much. This set-up is better than a power attenuator, for you can greatly overdrive the 60 watt amp without incurring high volume levels, (and the stress that it would create).

The pre-amp section is tube and uses switchable channels combined with master tone controls, so a guitarist can use a foot-pedal to switch from one clean channel to an overdriven channel and have the same volume level. The five band EQ is a passive tube unit which uses chokes as tone filters.

The Acoustic tube amps are interesting in that they have followed Fender and Boogie, and show how amps constantly evolve. It is also interesting to see a rather large company go back to tube electronics.

Fig. 123 Acoustic tube amp.

BURMAN

Burman is a relatively new company which could be described as the Mesa Boogie of the U.K. Whereas Boogie is built on Fender; Burman is built on Vox, Marshall, and other U.K. amps.

Burman's amps are all-tube and there are models from 70 watts to 140 watts — both as combo and piggyback units. For musicians fearing the demise of tube amps, Boogie and Burman must be truly auspicious signs.

The Pro 2000 combo is the main Burman model; it has 3-stage gain controls for shaping sustain, distortion, and clarity at any volume.

The Burman Company does not have a U.S. distributor, and the charges of shipping and duty are likely to keep them a U.K. experience.

Fig. 124 Burman 200 watt stack.

KELLY

Jim is a musician whose exacting demands led him to design amps. He prefers the softer sound of 6V6 tubes over 6L6 tubes. He believes the best sounding distortion is created in the output stage, not the pre-amp stage.

Fig. 125 Jim Kelly amps.

The EQ section of these amps uses active shelving circuits rather than passive resonant graphic EQ circuits. This means the Kelly amp has a wider range of tones than most resonate EQ circuits.

One nice feature Kelly offers is owner information that explains how to replace tubes, the best use of standby switches, effects loops, power attenuators, bias adjustment procedures, and trouble-shooting. This little guide would be useful to most tube amp owners, not just those with Kelly amps.

LEGEND

These hybrid amps, using a tube pre-amp and transistor, follow the principle that tube sound is developed primarily in the pre-amp, not the power amp section. Tube power amps are very expensive to manufacture since they require large power supplies, transformers, etc., and solid-state cuts costs considerably. Proponents of these hybrid amps believe it produces the best amps

Most Legend amps are styled after Boogie amps, i.e., the hardwood cabinet and cane grill. Legend amps do not sound like a Boogie amp because they are one-half transistor.

MUSICMAN

When Music Man first made amps in 1974, the company was headed by Forest White and Tom Walker. Leo Fender was involved with the product design. These amps have always been hybrid amps (solid state front and with tubes in the output section). The early models from 1974 to 1979 used 6CA7/EL34 type power tubes, but models after 1979 used 6L6GC tubes. These early models were available with alnico magnet speakers. After 1979 the 65 & 130 series amps went from one channel to a 2 channel, and the power on the 65 series boosted from 65 to 75 watts. The power on the 130 series was boosted from 130 to 150 watts.

The Music Man amp circuit is all solid-state except for the power tubes. These tubes are run at low levels and remain cool enough to touch. This greatly extends their life, but it does limit the tube's sound shaping effect. Even though Music Man amps use some tubes, they are considered more of a solid-state amp than a tube amp.

MESA BOOGIE

The Mesa Boogie Company was founded by Randall Smith in 1972. Earlier, in 1968, he was instrumental in the creation of Prune Music in the San Francisco bay area.

Randall has been working with electronics since he was ten, building and repairing tube ham radio sets. The first Boogie type amp was created when Randy took Barry (Country Joe Fish) Melton's Princeton Reverb apart and built a Fender 4-10 Bassman circuit using a JBL D120 speaker. Over the next few years, about 300 of these hot rodded Princetons were produced. Later, Boogie amps resembled a Twin Reverb more than a Bassman, i.e., 100 watts from four 6L6GC output tubes.

In 1972, Randy decided to build complete amps instead of rebuilding older ones and began producing the Mesa Boogie Mark I. The objective was to build the smallest possible tube amp capable of the qualities of high output, good sustain and good tone, regardless of the amp's volume.

Randy became interested in the fact that many guitarists liked a lot of sustain and he looked for ways to create it. He designed pre-amp circuits which would yield good-sounding overdrive. His purpose was to capture the best of an old 4-10 Bassman and blackface Twins.

When Randy brought out the Boogie Mark II, he had already build many Mark I's. He refined the circuit and added several features such as footpedal switchable channels, graphic equalizers, and half-power modes. The switchable channels allow for a choice between a clean sound and distortion sound. The latter is achieved by routing the output of channel one into the input of channel two. The graphic e.q. is a five band device that is an optional extra. The half-power mode occurs when two of the four output tubes are put on standby. This feature allows for the amp to be run at full volume for the utmost distortion and sustain, yet have a power output that is not too loud.

Randall has not stood still waiting for everyone to discover (or re-discover) tube amps. First there was the Mark I, then the Mark II, and now the Mark II-B. The basic Mark II-B has: two power levels (60 and 100 watts); four volume controls on the front panel; five pull-out control knobs controlling over-all Brightness, Gain Boost, Lead Drive Switch, Lead Brightness, and more. The amp also has a slave output jack and slave output level control, as well as effects send/receive points. And there is even more, a six-position power selector so the amp can be used with 100, 117, 200, 220, or 240 volt mains power. The Mark II-B also has the classic Boogie footswitchable clean and distortion channels. Note that Mesa Boogie makes many other configurations of amps in addition to the basic Mark II-B 100/60 with one 12" speaker. They make amp heads, speaker cabinets, and even a 60 watt model called Son of the Original Boogie.

Boogie amps started a move to smaller tube amps with custom features. In the years since, the Boogie was acclaimed as the world's greatest guitar amp. Many other amp companies have returned to their drawing boards and have produced tube amps with extra tonal circuits and wide operational flexibility. The Boogie can be held largely responsible for the rebirth of the tube amp.

Fig. 126 Mesa Boogie amps.

Fig. 127 More Boogie amps.

SUMMATION

Since this book is an introductory guide, there is not room for a fully detailed analysis of all amp companies and all the models made. However, it needs to be noted that there are more companies and models than presented in this book. For instance, older tube amps by Hi-Watt, Sound City, Sumn and Orange are thought of, by many, as very good examples of tube circuit design success. Also, it can be worthwhile to examine many smaller, older amps such as those made under the names of Harmony, Kay, Melody, or Kalamazoo (the last was an economy line of Gibson).

The best guide to checking out tube amps is to play a guitar through them and decide if you like the sound. A good extra measure, especially with very smaller or very old amps, is to plug the amp into a high quality speaker. Lastly, try to use the amp away from flourescent lights or light dimmers so that the normal hum level can be determined.

All technical information and historical dates have been checked and rechecked. In some instances, dates in this book may not coincide with what someone else may believe. Information was gleaned from employees (past and present) of amp companies, vintage amp dealers, amp catalogs, and sales brochures. There are some circumstances when it is hard to pinpoint events; i.e., an ad may predate or post-date a new amp. Employees may remember the amp in a factory in 1963, but no one saw it in stores until 1964. Though great effort was made, situations of "Was it late '57 or early '58?" do often occur.

A musician's choice of an amp is a personal matter, a bit like what type of guitar is "best". All in all, it is reasonable for a guitarist to own several amps, one tube oldie for its tone/distortion, a new tube model with all the effects available built into it, and a modern solid state amp that offers state-of-the-art design advances.

I hope you will enjoy this book as much as I have enjoyed writing it.

Donald Brosnac *Donald Brosnac*

INDEX

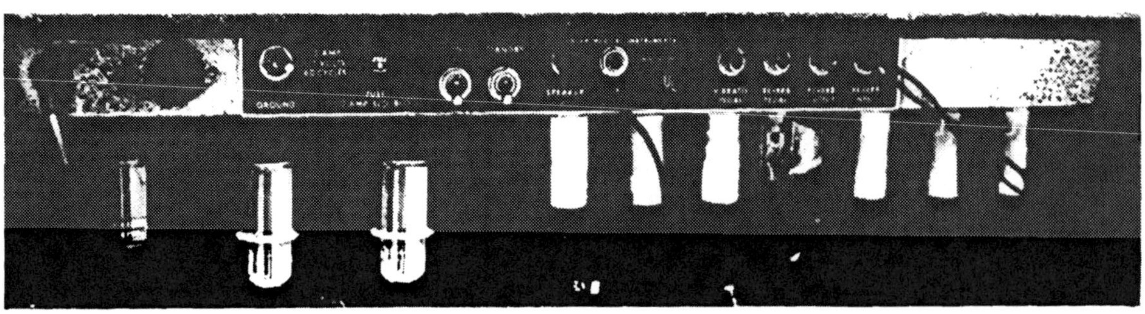

Fig. 128 Controls on back of typical Fender amp, from left to right: Ground polarity, Fuse, On-Off, Standby, Main speaker, Vibrato pedal, Reverb pedal, Reverb output, and Reverb input.

Other Titles Published or Distributed by The Bold Strummer Ltd.

Stock #	
I001	**ACOUSTICS & PERFORMANCE IN MUSIC** (Meyer) ISBN 0-933224-28-1 (cloth)
I002	**ADVENTURES IN TASTE** (Pohren) ISBN 0-933224-13-3 (cloth)
T052	**ALKAN v1: THE ENIGMA** (Smith) ISBN 0-912483-17-2 (cloth)
T053	**ALKAN v2: THE MUSIC** (Smith) ISBN 0-912483-44-X (cloth)
T054	**AMERICAN MINIMAL MUSIC** (Mertens) ISBN 0-912483-15-6 (paper)
T004	**AMP BOOK (THE)** (Brosnac) ISBN 0-933224-05-2 (paper)
T055	**ANATOMY OF A NY DEBUT RECITAL** (Montparker) ISBN 0-912483-48-2 (cloth)
T003	**ANIMAL MAGNETISM FOR MUSICIANS** (Zwaan) ISBN 0-933224-25-7 (paper)
T005	**ANTHOLOGY OF FLAMENCO FALSETTAS** (Mitchell) ISBN 0-933224-39-7 (paper)
T006	**ANTONIO DE TORRES - GUITAR MAKER** (Romanillos) ISBN 0-933224-92-3 (cloth)
T424	**ANTONIO DE TORRES - GUITAR MAKER** (Romanillos) ISBN 0-933224-93-1 (paper)
I007	**ART OF FLAMENCO (THE)** (Pohren) ISBN 0-933224-10-9 (paper)
T008	**ART OF PRACTICING (THE)** (Artzt) ISBN 0-933224-17-6 (paper)
T056	**AT THE PIANO WITH FAURE** (Long) ISBN 0-912483-38-5 (paper)
AIG002	**BACH: ADAGIO, ARIA & FUGUE** (Walker)
I098	**BELA BARTOK: ANALYSIS OF MUSIC** (Lendvai) ISBN 0-912483-33-4 (cloth)
I099	**BERNARD STEVENS & HIS MUSIC** (Stevens) ISBN 0-912483-20-2 (cloth)
AIG011	**BLUES IMPROMPTU** (Perry)
AIG016	**BRAHMS: THREE INTERMEZZI, OP. 117** (Walker)
I009	**BUILD YOUR OWN ELECTRIC GUITAR** (Foley) ISBN 0-961836-10-5 (paper)
I010	**BURNS BOOK (THE)** (Day) ISBN 0-933224-09-5 (paper)
T011	**CLASSIC GUITAR CONSTRUCTION** (Sloane) ISBN 0-933224-14-1 (paper)
AIG012	**CRUZANDO EL CHARCO** (Alegrias)
P075	**DENIS STEVENS DISCOGRAPHY (A)** ISBN 0-912483-57-1
T100	**DEVELOPMENT OF THE MODERN GUITAR** (Huber) ISBN 0-933224-59-1 (paper)
AIG001	**DOMENICO SCARLATTI: TEN SONATAS** (Willard)
AIG013	**EL DRON DEL FARAON**
I128	**EUROPEAN PIANO ATLAS** (Herzog) ISBN 0-933224-29-X (paper)
T012	**FENDER GUITAR (THE)** (Achard) ISBN 0-933224-48-6 (paper)
AIG010	**FIVE BAGATELLES** (DiPalma)
AIG114	**FLAMENCO CLÁSICO (CD)** (Kostner)
I013	**FLAMENCO GUITAR (THE)** (George) ISBN 0-933224-38-9 (cloth)
T302	**FLAMENCO OF CADIZ BAY** (Howson) ISBN 0-933224-72-9 (cloth)
T014	**GIBSON GUITAR FROM 1950 v1** (Bishop) ISBN 0-933224-46-X (paper)
T015	**GIBSON GUITAR FROM 1950 v2** (Bishop) ISBN 0-933224-47-8 (paper)
I129	**GLOSSARY OF HARPSICHORD TERMS** (Costa) ISBN 0-933224-30-3 (cloth)
I323	**GREAT PIANO VIRTUOSOS OF OUR TIME** (Lenz) ISBN 0-912483-73-3 (paper)
T416	**GREAT TUBE AMP & GUITAR MODS** (Thompson) ISBN 0-933224-90-7 (paper)
T134	**GUIDE TO PRACTICAL STUDY OF HARMONY** (Tchaikowsky) ISBN 0-912483-58-X (paper)
T020	**GUITAR FROM RENAISSANCE TO PRESENT DAY** (Turnbull) ISBN 0-933224-57-5 (paper)
T016	**GUITAR HISTORY v1: FENDER GUITARS** (Brosnac) ISBN 0-933224-06-0 (paper)
T017	**GUITAR HISTORY v2: GIBSON SGs** (Bulli) ISBN 0-933224-20-6 (paper)
T018	**GUITAR HISTORY v3: GIBSON CATALOGS** (Hetrick) ISBN 0-933224-56-7 (paper)

Stock #	Other Titles Published or Distributed by The Bold Strummer Ltd.
T303	**GUITAR HISTORY v4: THE VOX STORY** (Petersen & Denney) ISBN 0-933224-70-2 (paper)
T304	**GUITAR HISTORY v5: GUILD GUITARS** (Beesley) ISBN 0-933224-78-8 (paper)
T019	**GUITAR REPAIR** (Sloane) ISBN 0-933224-15-X (paper)
T046	**GUITAR TRADER GTR BULLETIN 1-1982** (Kummer) ISBN 0-933224-58-3 (paper)
T102	**GUITAR TRADER GTR BULLETIN 2-1983** (Kummer) ISBN 0-933224-58-3 (paper)
I411	**GYPSIES & FLAMENCO** (Leblon) ISBN 0-933224-71-5 (paper)
T400	**HIGH PERFORMANCE MARSHALL HANDBOOK** (Boehnlein) ISBN 0-933224-80-X (paper)
T021	**HISTORY & DEVELOPMENT OF AMERICAN GUITAR** (Achard) ISBN 0-933224-18-4 (paper)
T305	**HISTORY THROUGH THE OPERA GLASS** (Jellinik) ISBN 0-912483-90-3 (cloth)
AIG004	**HOMAGE** (Hand)
I022	**HOW TO SELECT A VIOLIN BOW** (Planta) ISBN 0-933224-31-1 (paper)
T023	**INTRO. TO SCIENTIFIC GUITAR DESIGN** (Brosnac) ISBN 0-933224-01-X (paper)
T069	**JANACEK: LEAVES FROM HIS LIFE** (Janacek & Tausky) ISBN 0-912483-32-6 (paper)
P066	**JOHN FOULDS & HIS MUSIC** (MacDonald) ISBN 0-912483-02-4 (paper)
T067	**JOY OF ORNAMENTATION (THE)** (Conforto & Stevens) ISBN 0-912483-49-0 (paper)
T068	**KENTNER: A SYMPOSIUM** (Taylor) ISBN 0-912483-11-3 (paper)
AIG008	**KEYS TO FLAMENCO, Vol. 1** (Kostner)
AIG009	**KEYS TO FLAMENCO, Vol. 2** (Kostner)
AIG014	**KEYS TO FLAMENCO, Vol. 3** (Kostner)
T024	**LEFT HANDED GUITAR** (Clarke) ISBN 0-933224-xx-x (paper)
T025	**LEFT HANDED GUITAR - CHART**
I135	**LIPATTI** (Tanasescu & Bargauanua/Grindea) ISBN 0-912483-18-0 (paper)
I136	**LISZT & HIS COUNTRY, 1869-1873** (Legany) ISBN 0-912483-27-X (cloth)
I026	**LIVES & LEGENDS OF FLAMENCO** 2/e (Pohren) ISBN 0-933224-12-5 (cloth)
ST330	**MAKING MASTER GUITARS** (Courtnal)
T027	**MAKING MUSICAL INSTRUMENTS** (Sloane) ISBN 0-933224-60-5 (paper)
I034	**MANUAL OF GUITAR TECHNOLOGY** (Jahnel) ISBN 0-933224-27-3 (cloth)
T320	**MANUEL M. PONCE & THE GUITAR** (Otero) ISBN 0-933224-84-2 (cloth)
T321	**MANUEL M. PONCE & THE GUITAR** (Otero) ISBN 0-933224-85-0 (paper)
T072	**MASCAGNI: AN AUTOBIOGRAPHY** (Stivender) ISBN 0-912483-06-7 (cloth)
T070	**MICHAEL TIPPETT O.M.: A CELEBRATION** (Lewis) ISBN 0-859361-40-3 (cloth)
T409	**MUSIC OF INDIA (THE)** (Massey & Massey) ISBN 0-912483-xx-x (paper)
T310	**MUSIC OF MY TIME (THE)** (Peyser) ISBN 0-912483-99-7 (cloth)
T109	**MUSIC OF SZYMANOWSKI (THE)** (Samson) ISBN 0-912483-32-2 (paper)
T035	**MUSICAL INSTRUMENT COLLECTOR** Rev/e (Willcut & Ball) ISBN 0-933224-00-1 (paper)
T096	**MUSICIAN'S GUIDE TO COPYRIGHT & PUBLISHING** (Wager) ISBN 0-912483-60-1 (paper)
T071	**MUSICIAN'S THEORY BOOK** (Hamerik) ISBN 0-912483-61-X (paper)
I111	**MY VIOLA & I** (Tertis) ISBN 0-912483-87-3 (paper)
T037	**NATURAL CLASSICAL GUITAR (THE)** (Ryan) ISBN 0-933224-40-0 (paper)
T112	**NUTLEY PAPERS (THE)** (Billings) ISBN 0-912483-88-1 (paper)
I077	**OPRICHNIK (THE)** (Tchiakovsky & Taylor) ISBN 0-912483-23-7 (spiral)
T141	**OPUS EST** (Rapoport) ISBN 0-912483-42-3 (paper)

Other Titles Published or Distributed by
The Bold Strummer Ltd.

Stock #	
I140	**PACO De LUCIA & FAMILY: The Master Plan** (Pohren) ISBN 0-933224-62-1
T405	**PACO De LUCIA: A New Tradition** (Sevilla)
I097	**PEACE SONGS** (Jordan) ISBN 0-912483-95-4 (paper)
AIG007	**PEDRO CORTÉS: THREE PIECES**
T113	**PERCUSSION INSTRUMENTS** (Blades) ISBN 0-933224-71-0 (cloth)
T114	**PERCUSSION INSTRUMENTS** (Blades) ISBN 0-933224-61-3 (paper)
T083	**PERCY GRAINGER: A PICTORIAL BIOGRAPHY** (Simon) ISBN 0-933224-64-4 (paper)
P093	**PERCY GRAINGER: A SOURCE GUIDE** (Lewis) ISBN 0-912483-56-3 (paper)
T408	**PERFORMING J.S. BACH'S KEYBOARD MUSIC** (Kochevitsky)
T081	**PIANIST'S TALENT (THE)** (Taylor) ISBN 0-912483-65-2 (paper)
I130	**PIANO & HOW TO CARE FOR IT** (Funke) ISBN 0-933224-32-X (paper)
I131	**PIANO HAMMER (THE)** (Pfeifer) ISBN 0-933224-33-8 (paper)
I132	**PIANO NOMENCLATURE** 2/e (Schimmel & Herzog) ISBN 0-933224-34-6 (paper)
T115	**PRO/AM BOOK OF MUSIC/MYTHOLOGY v1** (Lewis) ISBN 0-912483-51-2 (cloth)
T116	**PRO/AM BOOK OF MUSIC/MYTHOLOGY v2** (Lewis) ISBN 0-912483-82-2 (cloth)
T301	**PRO/AM BOOK OF MUSIC/MYTHOLOGY** (2 vol set) (cloth)
P079	**PRO/AM GUIDE US BOOKS MUSIC v1 '86** (Lewis) ISBN 0-912483-03-2 (paper)
P080	**PRO/AM GUIDE US BOOKS MUSIC v2 '87** (Lewis) ISBN 0-912483-14-4 (paper)
I082	**RAVEL ACCORDING TO RAVEL** (Perlemuter & Jourdan-Morhange) ISBN 0-912483-19-1 (cloth)
T145	**RAYMOND LEPPARD ON MUSIC** (Leppard & Lewis) ISBN 0-912483-96-2 (cloth)
T315	**RIOPLATENSE GUITAR (THE)** (Pinnell) ISBN 0-933224-42-7 (cloth)
T316	**RIOPLATENSE GUITAR (THE)** (Pinnell) ISBN 0-933224-43-5 (paper)
P118	**RONALD STEVENSON - Biography** (MacDonald) ISBN 0-912483-62-8 (paper)
T089	**SCHUBERT'S MUSIC PIANO 4-HANDS** (Weekley & Arganbright) ISBN 0-912483-55-5 (paper)
T038	**SEGOVIA TECHNIQUE (THE)** (Bobri) ISBN 0-933224-41-9 (paper)
AIG015	**16 PIECES FROM THE IRISH TRADITION** (Freeley)
	SKETCHES FROM MY LIFE
T137	**SOMETHING ABOUT THE MUSIC v2** (Lewis) ISBN 0-912483-66-0 (cloth)
T138	**SOMETHING ABOUT THE MUSIC v2** (Lewis) ISBN 0-912483-54-7 (paper)
AIG003	**SONATA PARA GUITARRA** (Zyman)
T039	**SOUND OF ROCK (THE)** (Doyle) ISBN 0-933224-19-2 (paper)
I150	**SPANISH GUITAR/GUITARRA ESPANOL** ISBN 0-933224-79-6 (paper)
T042	**STEEL STRING GUITAR (THE)** Updated/e (Young) ISBN 0-933224-08-7 (paper)
T040	**STEEL STRING GUITAR CONSTRUCTION** (Sloane) ISBN 0-933224-16-8 (cloth)
I043	**STEINWAY SERVICE MANUAL (THE)** (Mathias) {Re-issue 1997} ISBN 0-933224-84-8 (cloth)
T044	**STUDENT'S DICTIONARY MUSICAL TERMS (THE)** (paper) ISBN 0-933224-21-4
I407	**SYMPHONIES OF HAVERGAL BRIAN v1** (MacDonald) (paper)
I119	**SYMPHONIES OF HAVERGAL BRIAN v2** (MacDonald) ISBN 0-912483-36-9 (paper)
I120	**SYMPHONIES OF HAVERGAL BRIAN v3** (MacDonald) ISBN 0-900707-64-X (paper)
T139	**TENSIONS IN PERFORMANCE OF MUSIC** (Grindea) ISBN 0-912483-10-5 (paper)
ST401	**THINGS ABOUT THE GUITAR** (Ramirez) ISBN 0-933224-88-5 (paper)
I045	**TUNING OF MY HARPSICHORD (THE)** (Kellner) ISBN 0-933224-30-3 (paper)
T414	**VACUUM TUBE GUITAR & BASS AMP - Servicing** (Zottolla) ISBN 0-933224-97-4

Stock #	Other Titles Published or Distributed by The Bold Strummer Ltd.
T415	**VACUUM TUBE GUITAR & BASS AMP - Theory** (Zottolla) ISBN 0-933224-96-6
P140	**VERDI & WAGNER** (Lendvai) ISBN 0-912483-86-5 (cloth)
I122	**VILLA-LOBOS: THE MUSIC** (Peppercorn) ISBN 0-912483-36-9 (cloth)
I047	**VIOLIN: Precepts & Observations of A Luthier** (Arakelian) ISBN 0-933224-36-2 (paper)
I413	**VIOLIN MARKET (THE)** (Huber) ISBN 0-933224-65-6 (cloth)
I050	**WAY OF LIFE (A)** (Pohren) ISBN 0-933224-02-8 (cloth)
I049	**WAY OF LIFE (A)** (Pohren) ISBN 0-933224-03-6 (paper)
T318	**WIND CRIED (THE)** (Hecht) ISBN 0-933224-73-7 (cloth)
P142	**WORKS OF ALAN HOVHANESS: CATALOG** (Howard) ISBN 0-912483-00-8 (paper)
I121	**XENAKIS** (Matossian) ISBN 0-912483-35-0 (paper)
I141	**ZOLTAN KODALY: His Life In Pictures & Documents** (Eosze) ISBN 0-912483-29-6 (cloth)

CPSIA information can be obtained at www.ICGtesting.com
Printed in the USA
269786BV00003B/13-36/A

9 780933 224056